Statistical Physics

Daijiro Yoshioka

Statistical Physics

An Introduction

With 71 Figures and 7 Tables

Professor Daijiro Yoshioka
Department of Basic Science
The University of Tokyo
3-8-1 Komaba, Meguro
Tokyo, 153-8902
Japan
e-mail: *daijiro@toki.c.u-tokyo.ac.jp*

ISBN-10 3-540-28605-5 Springer Berlin Heidelberg New York
ISBN-13 978-3-540-28605-9 Springer Berlin Heidelberg New York

Library of Congress Control Number: 2006923850

Springer is part of Springer Science+Business Media

springer.com

Production and Typesetting: LE-TEX Jelonek, Schmidt & Vöckler GbR, Leipzig
Cover design: WMX Design GmbH, Heidelberg

SPIN 10[illegible] 57/3100YL - 5 4 3 2 1 0 Printed on acid-free paper

Preface

"More is different" is a famous aphorism of P.W. Anderson, who contributed rather a lot to the development of condensed-matter physics in the latter half of the 20th century. He claimed, by this aphorism, that macroscopic systems behave in a way that is qualitatively different from microscopic systems. Therefore, additional rules are needed to understand macroscopic systems, rules additional to the fundamental laws for individual atoms and molecules. An example is provided by the various kinds of phase transitions that occur. The state of a sample of matter changes drastically at a transition, and singular behavior is observed at the transition point. Another good example in which quantity brings about a qualitative difference is the brain. A brain consists of a macroscopic number of neural cells. It is believed that every brain cell functions like an element of a computer. However, even the most sophisticated computer consists of only a limited number of elements and has no consciousness. The study of the human brain is still developing.

On the other hand, the paradigm for macroscopic matter, namely thermodynamics and statistical physics, has a long history of investigation. The first and second laws of thermodynamics and the principle of equal probability in statistical physics have been established as laws that govern systems consisting of a macroscopic number of molecules, such as liquids, gases, and solids (metals, semiconductors, insulators, magnetic materials, etc.). These laws belong to a different hierarchy from the laws at the microscopic level, and cannot be deduced from the latter laws, i.e. quantum mechanics and the laws for forces. Therefore, a "theory of everything" is useless without these thermodynamic and statistical-mechanical laws in the real world. The purpose of this book is to explain these laws of the macroscopic level to undergraduate students who are learning statistical physics for the first time.

In this book, we start from a description of a macroscopic system. We then investigate ideal gases kinematically. Following on from the discussion of the results, we introduce the principle of equal probability. In the second and third chapters we explain the general principles of statistical physics on the basis of this principle. We start our discussion by defining entropy. Then temperature, pressure, free energy, etc. are derived from this entropy. This concludes Part I of the book. In Part II, from Chap. 4 onwards, we apply statistical physics to some simple examples. In the course of this application, we show that entropy, temperature, and pressure, when defined statistical-mechanically, coincide with the corresponding quantities defined thermodynamically. We consider only thermal-equilibrium states in this book. Most of our examples are simple systems in which interaction between particles is absent. Interaction, however, is essential for phase transitions. For an illustration of how a phase transition occurs, we consider a simple ferromagnetic system in Chap. 7. At this point, readers will be able to obtain a general idea about statistical physics: how a system in equilibrium is treated, and what can be known. In Part III, some slightly more advanced topics are treated. First, we consider first- and second-order phase transitions in Chaps. 8 and 9. Then, in Chap. 10, we return to our starting point of the ideal gas, and learn what happens at low temperature, when the density becomes higher.

Physics is one of the natural sciences, and the starting point of an investigation is the question "Why does nature behave like this?" Therefore, it is a good attitude to ask "why?" This question should be aimed only at natural phenomena, though. In this book, we give an explanation, for example, for various strange characteristics of rubber. However, it is often useless to ask "why?" about the methods used for solving these questions, or how an idea or concept used to treat a problem was obtained. For example, it is not fruitful to ask how the definition of entropy was derived. The expression for the entropy was obtained by a genius after trial and error, and it cannot be obtained as a consequence of logical deduction. Logical deduction can be done by a computer. Great discoveries in science are not things that can be deduced. They are rushes of ideas to the head. Some students stumble over these whys and hows of the methods, and fail to proceed. We hope that you will accept the various concepts that geniuses have introduced into science, and enjoy the beauty of the physics developed by the application of such concepts.

Tokyo,
October 2005 *Daijiro Yoshioka*

Contents

Part I General Principles

Symbols and Fundamental Constants

The symbols used in this book are listed alphabetically in Table 0.1. Fundamental constants are listed in Table 0.2.

Table 0.1. Symbols used in this book

Symbol	Name	Definition or meaning
C_P	Heat capacity at constant pressure	$T(\partial S/\partial T)_P$
C_V	Heat capacity at constant volume	$T(\partial S/\partial T)_V$
B	Magnetic field	$\mu_0(H+M)$
$D(E)$	Density of single-particle states	
E	Energy	Total energy of the system
E_F	Fermi energy	Chemical potential of fermion system at $T=0$
$F(T,V,N)$	Helmholtz free energy	$-k_\mathrm{B}T\ln Z$
$F_\mathrm{L}(T,V,N,\Psi)$	Landau free energy	Free energy for a given value of the order parameter
$f(E)$	Fermi distribution function	$1/[\mathrm{e}^{\beta(E-\mu)}+1]$
$G(T,P,N)$	Gibbs free energy	$-k_\mathrm{B}T\ln Y = F+PV$
H	Magnetic field	$(B/\mu_0)-M$
I	Moment of inertia	
J	Exchange interaction (Chaps. 7 and 9)	
$J(T,V,\mu)$	Grand free energy (Chap. 10)	$-k_\mathrm{B}T\ln \Xi = -PV$
M	Magnetization	Total magnetic moment per unit volume
N	Number of particles	
$n(E)$	Bose distribution function	$1/[\mathrm{e}^{\beta(E-\mu)}-1]$
P	Pressure	$T(\partial S/\partial V)_{E,N}$
$\boldsymbol{p}$	Momentum vector	
p_F	Fermi momentum	$\sqrt{2mE_\mathrm{F}}$

Table 0.1 continued

Symbol	Name	Definition or meaning
Q	Quantity of heat	
S	Entropy	$k_{\mathrm{B}} \ln W$
T	Temperature	$(\partial S/\partial E)_{V,N}^{-1}$
T_{c}	Critical temperature	
T_{F}	Fermi temperature	$E_{\mathrm{F}}/k_{\mathrm{B}}$
$U(S,V,N)$	Internal energy	$F + ST$
V	Volume	
W	Work	$\int F\,\mathrm{d}x$
$Y(T,P,N)$	Partition function for fixed T and P	
$Z(T,V,N)$	Partition function	$\sum_i \exp(-\beta E_i)$
α	Critical exponent of heat capacity	
β	Inverse temperature	$1/k_{\mathrm{B}}T$
β	Critical exponent of order parameter	
$\zeta(z)$	Riemann zeta function	$\sum_{n=1}^{\infty} n^{-z}$
λ_{T}	De Broglie wavelength	$h/\sqrt{2mk_{\mathrm{B}}T}$
μ	Chemical potential	$-T(\partial S/\partial N)_{E,V}$
μ	Atomic magnetic moment	
$\Xi(T,V,\mu)$	Grand partition function	
χ	Magnetic susceptibility	$\lim_{B\to 0}(\mu_0 M/B)$
Ψ	Order parameter	
$\Omega_0(E)$	Number of states	
$\Omega(E)$	Density of states	$\mathrm{d}\Omega_0/\mathrm{d}E$

Table 0.2. Fundamental constants. The numbers in parentheses indicate the standard uncertainty; for instance, $1.60217653(14)\times10^{-19}$ means that the most probable value is 1.60217653×10^{-19} and the standard uncertainty is 0.00000014×10^{-19}. The three constants c, ϵ_0, and μ_0 are defined to have the values listed here, and so there is no uncertainty in those values

Symbol	Value	Name
c	$299792458\,\mathrm{m\,s^{-1}}$	Velocity of light in vacuum
e	$1.60217653(14)\times10^{-19}\,\mathrm{C}$	Elementary charge
h	$6.6260693(11)\times10^{34}\,\mathrm{J\,s^{-1}}$	Planck constant
$\hbar$	$1.05457168(18)\times10^{34}\,\mathrm{J\,s^{-1}}$	Planck constant divided by 2π
k_{B}	$1.3806505(24)\times10^{-23}\,\mathrm{J\,K^{-1}}$	Boltzmann constant
m_{e}	$9.1093826(16)\times10^{-31}\,\mathrm{kg}$	Electron mass
m_{p}	$1.67262171(29)\times10^{-27}\,\mathrm{kg}$	Proton mass
N_{A}	$6.0221415(10)\times10^{23}\,\mathrm{mol^{-1}}$	Avogadro number
R	$8.1314472(15)\,\mathrm{J\,mol^{-1}\,K^{-1}}$	Gas constant
ϵ_0	$1/\mu_0 c^2$	Permittivity of free space
μ_0	$4\pi\times10^{-7}\,\mathrm{N\,A^{-2}}$	Permeability of free space

Part I

General Principles

1

Thermal Equilibrium and the Principle of Equal Probability

In this chapter, a brief introduction to thermal and statistical physics is given. We describe in general terms the systems to which this branch of physics can be applied, the relationship of thermal and statistical physics to other branches of physics, and the objectives of thermal and statistical physics. We consider the kinetic theory of gas molecules, and thereby introduce the essence of statistical physics.

1.1 Introduction to Thermal and Statistical Physics

It is well known that matter is constructed from atoms. The inner structure of atoms, of nuclei, and of nucleons has been clarified; the forces that act on these particles and the laws of motion for them are known. However, we cannot understand the properties of the pieces of matter of moderate size around us on the basis of these laws only. Ordinary matter consists of macroscopic numbers of atoms, ions, and electrons. For such macroscopic matter, knowing the rules of motion for individual particles is not enough to understand its properties. We need additional laws to understand the properties of gases, liquids, solids, and solid solutions of macroscopic size. Such additional laws are provided by thermal physics and statistical physics.

Thermal physics, or *thermodynamics*, treats a macroscopic sample of matter such as a gas or a solid as a black box. It provides general laws for the response of matter to actions from the environment. For instance, when we exert a force on a volume of gas in the form of a pressure, the gas will contract. When we give energy in the form of heat to a gas, it will either expand or increase its pressure. There are general relationships between these responses. Thermodynamics gives us such relationships. This branch of physics evolved from the necessity to increase the efficiency of the conversion of heat to work, which became important after the Industrial Revolution. The laws of thermodynamics are quite general; they are independent of the species of the atoms from which the matter is constructed, and independent of the interactions between the atoms. In fact, the laws of thermodynamics are not based on the fact that matter is made of atoms at all; they were established before the existence of atoms was verified. Without us knowing the microscopic laws governing atoms, it can still give an upper bound on the work that can be obtained from an engine, i.e. on the efficiency of the engine. It gives a general framework for every phenomenon related to heat, including living organisms, and thus it should play an important role in solving environmental problems.

In thermodynamics, the free energy, which is a function of the state variables such as the pressure, temperature, volume, and entropy, plays an important role. If the free energy is given, all properties of the thermal-equilibrium state can be known. However, in the framework of thermodynamics, the free energy can be known from measurement only; it cannot be obtained by theoretical calculation. *Statistical physics* is the branch of physics where a scheme to calculate the free energy is formulated. In statistical physics, we use the fact that matter consists of atoms. On the basis of a knowledge of the microscopic laws that govern the motion of atoms, and, most importantly, an additional law of statistical physics, statistical physics gives a general expression for the free energy. Since statistical physics starts from the microscopic level, it can discuss not only thermal-equilibrium states, but also nonequilibrium states. Small deviations from thermal equilibrium can be discussed by the use of linear response theory, and we can discuss such effects as electrical or thermal conductivity. However, the statistical physics of nonequilibrium states is not yet well established, especially for states far from equilibrium. It is an actively investigated branch of physics even today. In this book, we restrict our discussion mainly to equilibrium states.

1.2 Thermal Equilibrium

1.2.1 Description of a System in Equilibrium

What do you feel when you enter a room? You may feel cool or hot, you may feel motion of the air, or, if you happen to be an android equipped with a pressure sensor, you may be able to measure the pressure. If the room has been kept closed for a long time and if there is neither air conditioning nor an electric fan, you will not feel any motion of the air, and for a small room, the temperature will be the same everywhere in the room. Similarly, if you dip your finger into a cup that has been left untouched for a while after tea has been poured into it, you will be able to tell that the tea is still hot or has cooled down, but nothing else. As in these examples, a macroscopic volume of air or fluid left untouched for a while becomes motionless and reaches a state with a common temperature throughout. We call such motionless, uniform-temperature states *thermal-equilibrium states.*

An important thing is that states in thermal equilibrium are characterized by only a few pieces of information. When two cups of water have the same temperature, we cannot distinguish between those cups of water, regardless of the fact that the motions of the water molecules in the two cups are quite different. Although statistical physics is aimed at understanding the properties of liquids, solids, or gases on the basis of the laws of motion of atoms and molecules, the information necessary for us to describe and treat matter in thermal equilibrium is quite limited; what we need are only a few variables, known as *state variables*, such as the temperature, the pressure, the volume, or the concentration in the case of a solution.

It is true that in states of thermal nonequilibrium we need more variables, since the temperature may depend on the position, or there may be a flow in such a state, for instance. However, the number of variables needed is much much smaller than the total number of degrees of freedom of the motion of the molecules involved. We know that about 12 g of carbon (i.e. diamond or graphite) contains one mole of carbon atoms, namely $N_{\rm A} \simeq 6\times 10^{23}$ atoms.[1] In thermodynamics and statistical physics, we describe the state of matter using only a small number of state variables. Such a description is in accordance with our daily experience, and it is sufficient for utilizing matter.

1.2.2 State Variables, Work, and Heat

Let us elaborate a little more on state variables. Of the state variables, the volume, written as V in this book, is the easiest variable to understand. It is defined geometrically. When a sample of matter occupies a rectangular space whose three sides are equal to L_x, L_y, and L_z, its volume is given by $V = L_x \times L_y \times L_z$. The volume is measured in units of cubic meters (m^3) in the SI system. It is classified as an *extensive* state variable, since it is a property of the whole system and scales with the amount of matter. When two systems with the same volume are put together to form a single system, the volume doubles.

The *pressure*, written as P in this book, is defined mechanically. When a sample of matter of pressure P is bordered by a rectangular wall of area $S = L_x \times L_y$, the wall is pushed by the matter with a force $F = PS$. The pressure is measured in $\mathrm{N\,m^{-2}} = \mathrm{Pa}$ (pascals). It is classified as an *intensive* state variable, since it is defined in every part of the system and does not scale with the amount of matter. At least one of these two state variables can easily be assigned to the system as a boundary condition or constraint on the system.

A system reaches thermal equilibrium when it is isolated from its environment. In such a situation, the energy of the system must be conserved. Therefore, the total energy of the system E is a well-defined variable. When a system is in thermal equilibrium with its environment, its energy may fluctuate slightly. However, the average value $\langle E \rangle$ is still well defined. In some cases, the energy of the system contains mechanical energy of the center of

[1] $N_{\rm A} = 6.0221415 \times 10^{23}$ is called the Avogadro constant.

gravity. In a gravitational field, a system has gravitational potential energy. If the system is moving as a whole, or if it is observed by a moving observer, it has kinetic energy. These mechanical energies, which can easily be converted to other forms of energy, that is, they can be utilized easily, can be put aside in a discussion of thermal equilibrium. Therefore, we remove the mechanical energy of the center of mass from the total energy, and call the average of the rest of the energy the *internal energy* U. If there is no mechanical energy of the center of mass, the internal energy coincides with the average energy $\langle E \rangle$. This energy determines the properties of the system in equilibrium, and it is another extensive state variable.[2] By definition, U is the conserved energy E itself for an isolated system in the absence of mechanical energy. The unit for internal energy is the joule $\mathrm{J} = \mathrm{kg\,m^2\,s^{-2}}$. The actual value of the internal energy is difficult to measure, but it is not necessary to know it. The only thing that we need to know is the difference between the internal energies of two equilibrium states.

The amount of internal energy can be changed by performing *work* on the system or supplying *heat* to it. The work can be either mechanical or electromagnetic. Mechanical work can be done on a system by compressing the system. When the system is compressed sufficiently slowly that the system can be considered to be in equilibrium at all times, the pressure of the system is well defined during the compression. In this case the work W done on the system is given by

$$W = -\int_{V_1}^{V_2} P \, \mathrm{d}V \,, \tag{1.1}$$

where V_1 and V_2 are the initial and final values, respectively, of the volume. The minus sign on the right-hand side comes from the definition that the work is positive when the energy of the system increases.

Heat is a general term for energy transferred between systems that cannot be considered as work. For example, it can be given to a system by placing an electrical resistance R in the system and passing a current I through it. In this situation a quantity of heat $Q = RI \, \Delta t$ is given to the system in a time interval Δt because of Joule heating in the resistor. Heat and work can be given to a system by various means. They can also be extracted from a system. In all cases, the internal energy changes by an amount ΔU when a quantity of heat Q is supplied and work W is done on the system:

$$\Delta U = Q + W \,. \tag{1.2}$$

This equation is known as the *first law of thermodynamics.*

At this point it is important to realize that heat and work are not state variables. What we classify as state variables are those variables that are uniquely determined when the system is in thermal equilibrium with fixed boundary conditions. For instance, when the temperature and pressure of the

[2] Strictly speaking, the pressure P is also defined as an average, and it fluctuates slightly in space and time as U does.

environment are fixed, a system in equilibrium has a certain pressure and internal energy, which can be observed or measured in principle. However, we cannot tell what is the amount of heat and work in the system. In fact, there is no heat or work in thermal equilibrium; only the internal energy exists in that situation. It is possible to extract all the internal energy as heat. However, this does not mean that there is a quantity of heat Q equal to the internal energy U. This is because we can also convert part of the internal energy to work, and utilize it. The amount of work extracted depends on the process. So quantities of heat and work are not uniquely determined in thermal equilibrium, and therefore they cannot be classified as state variables. We should not consider these variables to exist in thermal equilibrium. Heat and work appear only when the system changes from one equilibrium state to another.

1.2.3 Temperature and the Zeroth Law of Thermodynamics

A system with less internal energy is felt to be cool, and the same system with more internal energy is felt to be hot. The *temperature* is an intensive state variable used to describe quantitatively how cool or how hot a system is. For a given system, the temperature is an increasing function of the internal energy. The value of the temperature needs to be determined in such a way that two systems with the same temperature are felt by us to have the same hotness. For that purpose, two systems in mutual thermal equilibrium are defined to have the same temperature. We mean by mutual thermal equilibrium the state of two systems kept in contact with each other for a long time so that energy can be transferred between the systems, and the two systems have reached an equilibrium state. It has been established, as the *zeroth law of thermodynamics*, that this definition of equal temperature is consistent. Suppose that two systems A and B are in mutual thermal equilibrium and have the same temperature. Then, if another system C has the same temperature as A, this law guarantees that C has the same temperature as B. From this experimental fact, we can prepare a system to be used as a thermometer. Any system can be used as a thermometer if there is a visible change in the system as the temperature of the system changes, and there is a one-to-one correspondence between the temperature and appearance of the system. When such a system A is brought into thermal equilibrium with system B or C, the temperature of system A that is known from the appearance of the system gives the temperature of system B or C. The zeroth law guarantees that when B and C have been measured to have the same temperature by means of A, they are in mutual thermal equilibrium.

Since the temperature of a system is an increasing function of its internal energy, one way to define a temperature scale is to make the temperature a linear function of the internal energy of some standard sample of matter. Historically, temperature was defined by means of this concept. The freezing point and boiling point of water under ambient pressure were defined to be

0 and 100 degrees Celsius, respectively, and temperatures between these two fiducial points were determined such that the temperature became a linear function of the internal energy. It was defined that the temperature of one gram of water was raised by one degree Celsius when a quantity of heat of about 4.2 J was supplied to the sample.[3] However, this definition that depends on a specific material has been replaced by a material-independent definition of the absolute temperature, which we shall explain later. The unit in which absolute temperature is measured is the kelvin (K). A temperature t °C corresponds to an absolute temperature $T = t + 273.15\,\mathrm{K}$.

It should be noted that, for the definition of equal temperature, we have used the experimental fact that two systems with different temperatures reach a state of mutual thermal equilibrium when in energetic contact. During the course of the approach to equilibrium, energy in the form of heat moves from the system at higher temperature to the other system, so the equilibrium temperature attained is somewhere between the two initial temperatures. The fact that heat will not be transferred in the other direction without the intervention of a third system is known as the *second law of thermodynamics*.

1.2.4 Heat Capacity and Specific Heat

We have stated that temperature was historically defined to have a linear relationship with the internal energy of water. However, the precise definition of temperature is slightly different, and the internal energy of a general system, including one consisting of water, is not a linear function of the absolute temperature. How the internal energy of a system changes as a function of temperature depends on the particular system, and the derivative of the energy with respect to temperature is related to a quantity called the *heat capacity*. More precisely, the heat capacity is defined as the ratio of a quantity of heat ΔQ and a temperature difference ΔT in the limit $\Delta T \to 0$, where ΔQ is the quantity of heat needed to raise the temperature of the system by ΔT. Namely, the heat capacity C is defined by

$$C = \lim_{\Delta T \to 0} \frac{\Delta Q}{\Delta T} . \tag{1.3}$$

In fact, the heat capacity depends on the boundary condition of the system. When heat is supplied keeping the volume of the system constant, the heat capacity is called the *constant-volume heat capacity*, written as C_{V}. In this case, work related to a volume change is not done. In the absence of any other forms of work, the increase in the internal energy ΔU is equal to ΔQ, and thus

$$C_{\mathrm{V}} = \lim_{\Delta T \to 0} \frac{\Delta U}{\Delta T} = \left(\frac{\partial U(T,V)}{\partial T} \right)_V . \tag{1.4}$$

[3] In the old system of measurement, a quantity of heat of 4.18605 J was called 1 calorie.

The subscript V on the right-hand side of this equation indicates explicitly that the partial derivative should be taken with fixed V. This is a standard notation in thermodynamics.

On the other hand, when heat is supplied while the system is kept in a constant-pressure environment, thermal expansion of the system extracts work from the system to the environment, and so more heat is needed to raise the temperature of the system. The heat capacity in this case is called the *constant-pressure heat capacity*, written as C_{P}. From (1.2), $\Delta Q = \Delta U + P\Delta V$, and thus

$$\begin{aligned} C_{\mathrm{P}} &= \left(\frac{\partial U(T,V)}{\partial T}\right)_P + P\left(\frac{\partial V}{\partial T}\right)_P \\ &= \left(\frac{\partial U(T,V)}{\partial T}\right)_V + \left(\frac{\partial U(T,V)}{\partial V}\right)_T \left(\frac{\partial V}{\partial T}\right)_P + P\left(\frac{\partial V}{\partial T}\right)_P \\ &= C_{\mathrm{V}} + \left\{\left(\frac{\partial U(T,V)}{\partial V}\right)_T + P\right\}\left(\frac{\partial V}{\partial T}\right)_P . \end{aligned} \tag{1.5}$$

The heat capacity is an extensive quantity; the heat capacity per unit mass or per mole is called the *specific heat* or *molar heat*, respectively. The specific heat of a dilute gas is almost temperature-independent, and so the internal energy of a sample of a gas is proportional to the temperature. On the other hand, the specific heat of a solid decreases at low temperature. We shall discuss these behaviors in the following chapters.

1.3 Kinetic Theory of Gas Molecules

Why is it all right to use only a few state variables to describe the state of matter in thermal equilibrium? Knowing the answer to this question leads directly to the essence of statistical physics. To answer this question, we start by considering an ideal gas from the standpoint of kinetic theory. The air around us has an almost uniform density and temperature on a scale of a few meters. The pressure P and the temperature T obey approximately the equation of state of an ideal gas, namely the Boyle–Charles law,

$$PV = nRT . \tag{1.6}$$

Here, n is the quantity of gas in the volume V measured in moles, and $R = 8.314472\,\mathrm{J\,mol^{-1}\,K^{-1}}$ is the gas constant. Let us see how this property arises from the law of motion for gas molecules.

If we think naively, the gas molecules must cooperate to form such a state of uniform density and uniform temperature. However, this expectation is not true. We shall see that this result actually comes from the fact that each gas molecule behaves independently of the others. The molecules can move freely with any velocity they like. Nevertheless, the Boyle–Charles law is obtained.

1.3.1 The Spatial Distribution of Gas Molecules

Average Number of Molecules

First, we shall see that a uniform density is obtained even though each molecule moves independently. There are several assumptions that we need to obtain the results presented below:

- A gas consists of molecules.
- The number of molecules in the volume that we are considering is enormous. For example, in the standard state, there are approximately 2.7×10^{22} molecules in one liter of air.
- The gas molecules do not interact with each other.

The last assumption is an idealization, and so we are going to consider an ideal gas. Strictly speaking, there are weak interactions between molecules in real gases. In addition to these assumptions, we restrict ourselves to the case in which the gas consists of a single species of molecules.

Since there are a huge number of molecules moving independently and we cannot know the position or velocity of each molecule, what we discuss is the probability distribution of the molecules. Since we cannot observe the position of a molecule, we assume that it can be anywhere in a container of volume V. This volume V is one of constraints we can assign to the gas. We put N molecules into a box of this volume, and examine the probability distribution of the number of molecules in a small volume V_1 in the box as shown in Fig. 1.1. The fact that the density of the air is uniform means that the number of molecules in the volume V_1 is NV_1/V. However, since the molecules are moving independently, there should be temporal fluctuations in this number. We examine this fluctuation on the basis of the probability p for one molecule to be found in the volume V_1.[4] It is natural to put $p = V_1/V$.

Let us write the probability of finding n molecules in V_1 as $W_N(n)$. This probability is given as follows:

$$W_N(n) = \frac{N!}{n!(N-n)!} p^n q^{N-n} , \tag{1.7}$$

where $q = 1 - p$. In this expression, the first factor is the number of ways in which n molecules can be taken from N, p^n is the probability that these

[4] We neglect gravity here. A gravitational field makes the probability slightly higher when V_1 is placed near the floor of the box V than when it is placed near the ceiling. However, this effect is negligibly small when we consider a system whose height is of the order of a meter.

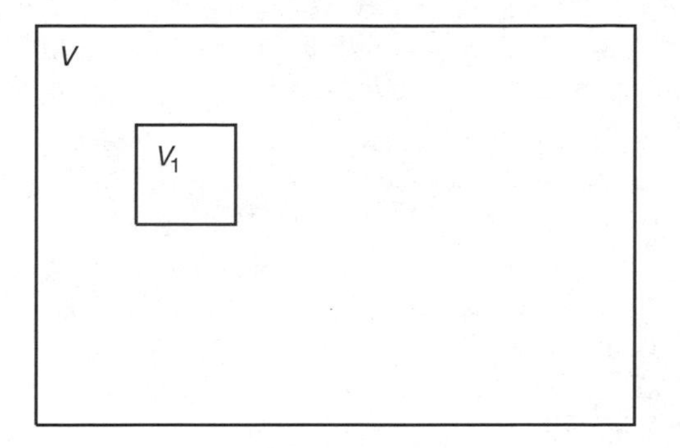

Fig. 1.1. A volume V and a part of it, V_1. We put N molecules in the volume V, and guess the number of molecules in V_1

n molecules are in V_1, and q^{N-n} is the probability that the other $N-n$ molecules are somewhere outside of the volume V_1. From this probability, the expectation value of the number of molecules in V_1, which we denote by $\langle n \rangle$, is given by

$$\begin{aligned}\langle n \rangle &= \sum_{n=1}^{N} n W_N(n) \\ &= \sum_{n=2}^{N} Np \frac{(N-1)!}{(n-1)!(N-n)!} p^{n-1} q^{N-n} \\ &= Np(p+q)^{N-1} = Np \\ &= N\frac{V_1}{V}, \end{aligned} \tag{1.8}$$

as expected. Here we have used the binomial theorem, described in Appendix A.

Variance of the Distribution

Next we calculate the mean square deviation, or variance, of this probability distribution. This is calculated as follows:

$$\begin{aligned}\langle (n - \langle n \rangle)^2 \rangle &= \langle n^2 - 2n\langle n \rangle + \langle n \rangle^2 \rangle \\ &= \langle n^2 \rangle - \langle n \rangle^2 \\ &= \langle n(n-1) \rangle + \langle n \rangle - \langle n \rangle^2 \\ &= \sum_n n(n-1) W_n(n) + Np - N^2p^2 \\ &= N(N-1)p^2 + NP - N^2p^2 \\ &= Npq\,. \end{aligned} \tag{1.9}$$

In the course of this calculation, we have used the binomial theorem again. From this variance, the relative deviation is

$$\frac{\sqrt{\langle (n - \langle n \rangle)^2 \rangle}}{\langle n \rangle} = \frac{1}{\sqrt{N}} \sqrt{\frac{q}{p}} \ll 1 . \tag{1.10}$$

This deviation is unbelievably small. Let us see how small this value is. Consider a sample of gas consisting of a macroscopic number of molecules. For example, there are about 2.7×10^{25} molecules in a volume of $1\,\mathrm{m}^3$ of air in the standard state. Let us take $p = 0.1$, or $V_1 = 0.1\,\mathrm{m}^3$. Then $\langle n \rangle = 2.7 \times 10^{24}$, and the standard deviation $\sqrt{Npq} \simeq 10^{12}$. So the number fluctuates by $\pm 10^{12}$ around the mean value. This fluctuation itself is quite large. However, it is negligibly small compared with the mean value, $\simeq 10^{24}$. To realize how small

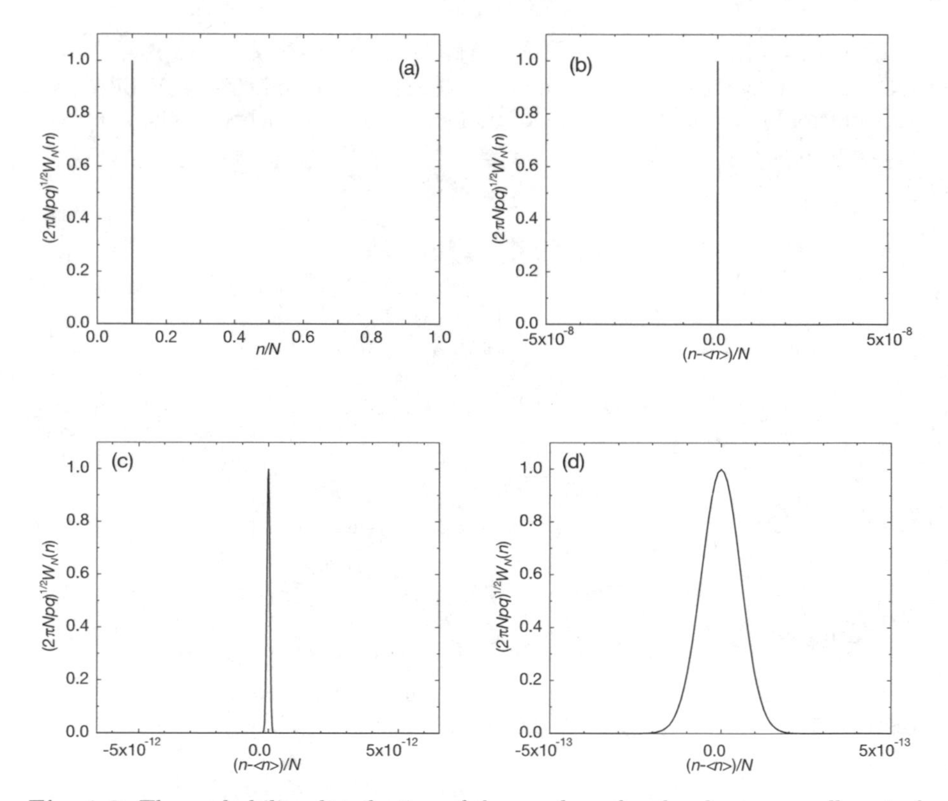

Fig. 1.2. The probability distribution of the number of molecules in a small part of a system. The probability of finding n molecules in a volume $V_1 = 0.1\,\mathrm{m}^3$, $W_N(n)$, when there are 2.7×10^{25} molecules in $V = 1\,\mathrm{m}^3$ is plotted. The *vertical axis* shows $\sqrt{2\pi Npq} W_N(n)$, normalized such that the peak value is approximately unity. (**a**) The whole range of possible variation $0 < n < N$ is shown. (**b**) The *horizontal axis* has been expanded 10 million times, and only the region around the peak is shown. (**c**) The *horizontal axis* has been expanded 7600 times further. (**d**) The *horizontal axis* has been expanded 10^5 times from the situation shown in (**b**)

it is, let us try to plot the probability distribution, taking n as the horizontal axis and the probability $W_N(n)$ as the vertical axis.

Figure 1.2a shows the case where the whole range of n is plotted on a sheet of ordinary size. The peak is at one-tenth of the length of the whole horizontal axis from the origin. The width of the distribution function is much smaller than the width of the line; in fact, it is smaller than the size of a nucleus. Next we imagine a larger sheet of paper 500 km wide, almost the distance between Berlin and Vienna, or 1.5 times the distance between Paris and London. The peak will then be somewhere between those cities. Figure 1.2b shows what is plotted around the peak. The width of this peak is about 0.1 μm, and is still too small to be seen. Finally, let us imagine a sheet of width 3.8 million kilometers, 10 times the distance between the earth and the moon. Then the peak is situated on the surface of the moon. There, an astronaut can see a distribution like that in Fig. 1.2c. Now we can see the width of the peak. Figure 1.2d shows the shape of the peak expanded even further. You should imagine the meaning of the size of the peak in Fig. 1.2c compared with the vast distance to the moon. This sharpness of the peak of the distribution is the essence of statistical physics. A sample of matter of ordinary size consists of a macroscopic number of atoms or molecules. Because of this fact, the probability distribution of every observable quantity shows a very sharp peak around the mean value, and the mean value coincides with the peak of the distribution, as we shall see next.

Details of the Distribution

We can rewrite the probability distribution $W_N(n)$ around the peak using the fact that N and n are macroscopic numbers. For macroscopic numbers, the factorial function can be approximated by Stirling's formula,[5]

$$\ln N! \simeq N \ln N - N \quad (N \gg 1)\,. \tag{1.11}$$

Then we obtain

$$\begin{aligned} W_N(n) &\simeq \frac{\mathrm{e}^{N\ln N - N}}{\mathrm{e}^{n\ln n - n}\mathrm{e}^{(N-n)\ln(N-n)-N+n}} p^n q^{N-n} \\ &= \mathrm{e}^{N\ln N - n\ln n - (N-n)\ln(N-n)} p^n q^{N-n} \\ &= \mathrm{e}^{n\ln(N/n)+(N-n)\ln[N/(N-n)]} p^n q^{N-n} \\ &= \left(p\frac{N}{n}\right)^n \left(q\frac{N}{N-n}\right)^{N-n} . \end{aligned} \tag{1.12}$$

[5] See Sect. A.2 for Stirling's formula. We use ln for the natural logarithm $\log_{\mathrm{e}}$ in this book.

The position of the peak of $W_N(n)$ is determined by the condition

$$\begin{aligned} \frac{\mathrm{d}}{\mathrm{d}n} \ln W_N(n) &\simeq \frac{\mathrm{d}}{\mathrm{d}n} \left\{ n \ln \left(p \frac{N}{n} \right) + (N-n) \ln \left(q \frac{N}{N-n} \right) \right\} \\ &= \ln \left(\frac{pN}{n} \right) - 1 - \ln \left(q \frac{N}{N-n} \right) + 1 \\ &= 0 \,. \end{aligned} \tag{1.13}$$

Namely,

$$\frac{p}{n} = \frac{q}{N-n} \,, \tag{1.14}$$

or

$$Np = (p+q)n = n \,. \tag{1.15}$$

Therefore the peak is positioned at $n = \langle n \rangle$.

Furthermore,

$$\begin{aligned} \frac{\mathrm{d}^2}{\mathrm{d}n^2} \ln W_N(n) &\simeq \frac{\mathrm{d}}{\mathrm{d}n} \left\{ \ln \left(\frac{pN}{n} \right) - \ln \left(\frac{qN}{N-n} \right) \right\} \\ &= -\frac{1}{n} - \frac{1}{N-n} = -\frac{1}{Np} - \frac{1}{Nq} \\ &= -\frac{1}{Npq} \,. \end{aligned} \tag{1.16}$$

Therefore, the Taylor expansion of $\ln W_N(n)$ around the peak is

$$\ln W_N(n) = \ln W_N(\langle n \rangle) - \frac{1}{2} \frac{1}{Npq} (n - \langle n \rangle)^2 + O[(n - \langle n \rangle)^3] \,. \tag{1.17}$$

That is, the distribution function is a Gaussian around the peak:[6]

$$W_N(n) \simeq W_N(\langle n \rangle) \exp \left[-\frac{1}{2} \frac{1}{Npq} (n - \langle n \rangle)^2 \right] \,. \tag{1.18}$$

In summary, we have found the following for the probability distribution of gas molecules:

- The distribution has a peak at the mean value.
- There is a fluctuation around the peak.
- However, the fluctuation is so small that a nonuniform distribution is not observed in practice.

We can imagine that these properties will not be restricted to the spatial distribution of gas molecules, but may be possessed by any distribution in thermal equilibrium in which a macroscopic number of molecules are involved.

---

[6] For details of the Gaussian distribution function, see Appendix B. The result (1.17) is an example of the central limit theorem, described in Sect. B.1.

1.3.2 Velocity Distribution of an Ideal Gas

In the previous subsection, we simply assumed that each molecule can be anywhere in the whole volume with uniform probability. Then we recovered the experimental fact that the density of a gas is uniform in thermal equilibrium. This assumption is one example of the *principle of equal probability*, which we shall describe later. In this subsection, we consider the velocity distribution on the basis of a similar assumption. The molecules change their velocities owing to collisions with each other and with the walls of the container. We cannot observe the velocity of every molecule; even if it were possible to know it, it would be hopeless to attempt to record all the velocities of the molecules. This situation is similar to that for the observation of the position of every molecule. Thus, it might be a good idea to assume that every molecule can have any velocity it likes, just as it can be anywhere in the container. Someone once thought that this should work, since the number of molecules is macroscopic. You may think that this idea may be too optimistic. However, to disclose the conclusion first, the idea worked! Moreover, this kind of idea has been generalized into the principle of equal probability, which states that "at the microscopic level, every possible state is realized with the same probability". The statistical physics of the equilibrium state is constructed upon this principle, and every consequence of this theory coincides with experimental results.

The construction of statistical physics on the basis of this principle is described in the following chapters. Here, we return to the problem of the

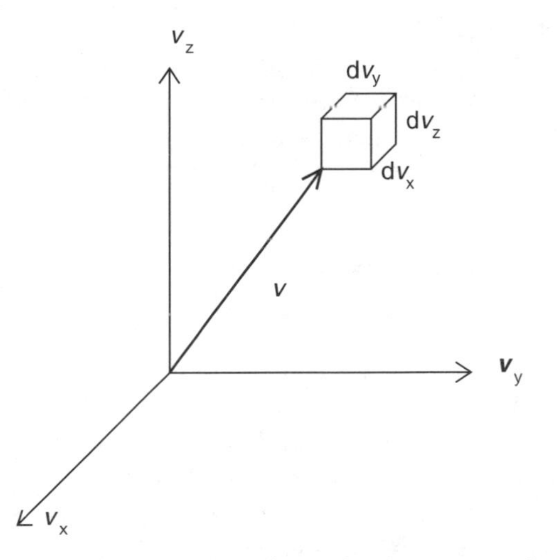

Fig. 1.3. Velocity space. The velocity of a molecule is mapped to a point in this space. The velocity distribution of all the molecules is described by a distribution of points representing each molecule in this space. The distribution is specified by the numbers of molecules in infinitesimally small boxes of volume $dv_x\,dv_y\,dv_z$ around each velocity $\boldsymbol{v}$

velocity distribution. In Sect. 1.3.1, the molecules must observe the constraint that they should be in a container of volume V. In the present case of the velocity distribution, there is an additional constraint that the total energy of the molecules E should be a constant. These constraints, specified by V and E, characterize the system when it is in thermal equilibrium. These constraints are nothing but the state variables with which we describe the system in equilibrium.

For the following discussion, we need to introduce velocity space, which is shown in Fig. 1.3. The velocity of a molecule is mapped to a point in this space. The fact that each molecule can take any velocity means that the point representing the velocity of a molecule can be anywhere in this infinite space with equal probability. Thus, if we divide this space into small boxes of volume $\mathrm{d}v_x\,\mathrm{d}v_y\,\mathrm{d}v_z$, the probability g of finding a molecule in this space is proportional to this volume, and $g/(\mathrm{d}v_x\,\mathrm{d}v_y\,\mathrm{d}v_z)$ is independent of the position of the small box. Each molecule is distributed with this probability in this space, and the velocity distribution of all the molecules is described by the numbers of molecules in all of the small boxes. Now we calculate a probability for a particular distribution where the ith box has n_i molecules, namely where the first box has n_1 molecules, the second box has n_2 molecules, and so on. This probability is given by

$$\begin{aligned} W(n_1, n_2, \ldots) &\propto \frac{N!}{n_1! n_2! \cdots n_i! \cdots} g_1^{n_1} g_2^{n_2} \cdots \\ &\simeq \left(\frac{N}{n_1}\right)^{n_1} \left(\frac{N}{n_2}\right)^{n_2} \cdots g_1^{n_1} g_2^{n_2} \cdots . \end{aligned} \tag{1.19}$$

Here g_i is the probability for a molecule to be in the ith box, and is proportional to the volume of the box. Stirling's formula, $N! \simeq N^N \mathrm{e}^{-N}$ $(N \gg 1)$, has been used to derive the final form.

Now the total number of molecules N and the total energy E are expressed as follows:

$$N = n_1 + n_2 + \cdots , \tag{1.20}$$

$$E = n_1 \epsilon_1 + n_2 \epsilon_2 + \cdots , \tag{1.21}$$

where $\epsilon_i = (m/2)\boldsymbol{v}_i{}^2$ is the kinetic energy of a molecule in the ith box, whose center is situated at $\boldsymbol{v}_i$. The maximum probability W under the constraints of fixed N and E is the most probable velocity distribution. It should be almost the unique distribution in thermal equilibrium, just as we have seen for the spatial distribution of the molecules, because of the macroscopic number of molecules involved. To find the maximum of W under the constraints, we apply Lagrange's method of undetermined multipliers to $\ln W$.[7] The conditions for the maximum are obtained as follows. Introducing λ and β as undeter-

[7] Lagrange's method of undetermined multipliers is described briefly in Appendix C.

mined multipliers, we require the following equations to be satisfied for each n_j, $j = 1, 2, 3, \cdots$:

$$\frac{\partial}{\partial n_j}\left(\ln W - \lambda \sum_i n_i - \beta \sum_i n_i \epsilon_i\right) = 0 \,. \tag{1.22}$$

Differentiating, we obtain

$$0 = \ln N - \ln n_j - 1 + \ln g_j - \lambda - \beta \epsilon_j \,. \tag{1.23}$$

Thus, the value of n_j that gives the maximum probability is

$$n_j = N \exp(-\lambda - 1 - \beta\epsilon_j) g_j \,. \tag{1.24}$$

The undetermined multipliers λ and β are determined so as to fulfill the following equations:

$$N = \sum_j N \exp(-\lambda - 1 - \beta\epsilon_j) g_j \tag{1.25}$$

and

$$E = \sum_j \epsilon_j N \exp(-\lambda - 1 - \beta\epsilon_j) g_j \,. \tag{1.26}$$

These summations are performed as integrals over velocity space. Since g_j is proportional to the volume of the box $\mathrm{d}v_x\,\mathrm{d}v_y\,\mathrm{d}v_z$, we simply put $g_j = c\,\mathrm{d}v_x\,\mathrm{d}v_y\,\mathrm{d}v_z$. Noticing that $\epsilon_j = (m/2)\boldsymbol{v}_j^2 = (m/2)(v_{jx}^2 + v_{jy}^2 + v_{jz}^2)$, we calculate (1.25) as follows:[8]

$$\begin{aligned} N &= \int_{-\infty}^{\infty}\int_{-\infty}^{\infty}\int_{-\infty}^{\infty} N \exp(-\lambda - 1) \exp\left[-\frac{m\beta}{2}(v_x^2 + v_y^2 + v_z^2)\right] c\,\mathrm{d}v_x\,\mathrm{d}v_y\,\mathrm{d}v_z \\ &= N\left(\frac{2\pi}{m\beta}\right)^{3/2} c \exp(-\lambda - 1) \,. \end{aligned} \tag{1.27}$$

This gives $c\exp(-1-\lambda) = (m\beta/2\pi)^{3/2}$. We input this result into (1.26) to obtain E:

$$\begin{aligned} E &= N\left(\frac{m\beta}{2\pi}\right)^{3/2} \int_{-\infty}^{\infty}\int_{-\infty}^{\infty}\int_{-\infty}^{\infty} \left(\frac{m}{2}\boldsymbol{v}^2\right) \exp\left(-\frac{m\beta}{2}\boldsymbol{v}^2\right) \mathrm{d}v_x\,\mathrm{d}v_y\,\mathrm{d}v_z \\ &= \frac{3N}{2\beta} \,. \end{aligned} \tag{1.28}$$

That is, $\beta = 3N/2E$. This β turns out to be equal to $1/k_{\mathrm{B}}T$, as shown later, where $k_{\mathrm{B}} = 1.380650 \times 10^{-23}\,\mathrm{J\,K^{-1}}$ is called the Boltzmann constant, and is equal to the gas constant R divided by the Avogadro constant N_{A}.

---

[8] For integrals of this kind, see Appendix B.

As a result of these calculations, we find finally that the most probable number of molecules in a box of volume $\mathrm{d}v_x\,\mathrm{d}v_y\,\mathrm{d}v_z$ at $\boldsymbol{v}$ in velocity space is

$$n(\boldsymbol{v})\,\mathrm{d}v_x\,\mathrm{d}v_y\,\mathrm{d}v_z = N\left(\frac{m\beta}{2\pi}\right)^{3/2}\exp\left(-\frac{m\beta}{2}v^2\right)\mathrm{d}v_x\,\mathrm{d}v_y\,\mathrm{d}v_z\,. \tag{1.29}$$

This most probable distribution is almost always realized in thermal equilibrium, and is called the *Maxwell distribution*. This distribution has been experimentally confirmed in thermal equilibrium. It has a peak at $\boldsymbol{v} = 0$. However, the number of molecules whose absolute value of velocity lies between v and $v + \mathrm{d}v$ is given by $4\pi v^2 n(v)\,\mathrm{d}v = 4\pi N(m\beta/2\pi)^{3/2}v^2\,\exp(-m\beta v^2/2)\,\mathrm{d}v$, and this distribution of speed is peaked at $v = \sqrt{2/m\beta}$ as shown in Fig. 1.4.

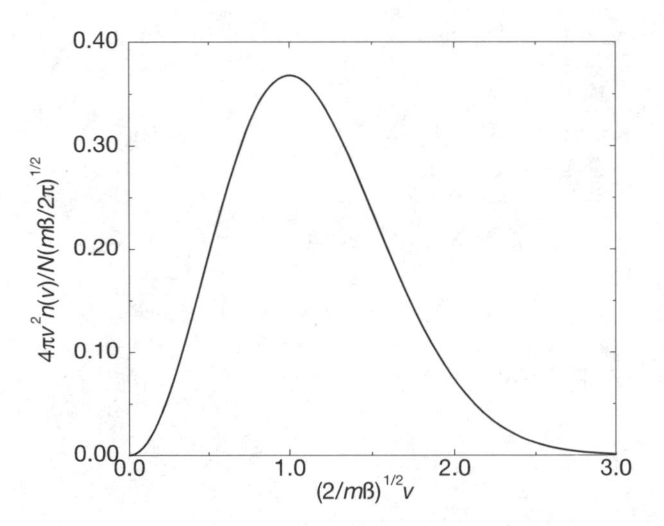

Fig. 1.4. The distribution of the speed of molecules, $4\pi v^2 n(v)$. To make the figure dimensionless, $4\pi v^2 n(v)$ divided by $N\sqrt{m\beta/2\pi}$ has been plotted as a function of $\sqrt{2/m\beta}v$. The peak is attained at $v = \sqrt{2/m\beta}$, or when the kinetic energy of a molecule is equal to $1/\beta(= k_\mathrm{B}T)$

1.3.3 The Pressure of a Gas

Gas molecules enclosed in a container of volume V collide with the walls of the container repeatedly. These collisions are the origin of the pressure. The pressure is independent of the material from which the walls are made.[9] Therefore, for the calculation of the pressure, we can assume that the

[9] If the pressure depended on the material, we could make a yacht that sailed without any wind, if we were to make a sail for the yacht where the two sides were covered with different materials. The second law of thermodynamics tells us that such a sail cannot be realized.

walls are perfectly flat and are so hard that a collision is a perfect elastic reflection.

Let us assume that the gas is contained in a rectangular box whose three sides have lengths L_x, L_y, and L_z as shown in Fig. 1.5. We calculate the contribution of the ith molecule to the pressure on the wall that is perpendicular to the x-axis. This molecule has a velocity whose x-component is v_{ix}. In a collision, this molecule gives an impulse $2mv_{ix}$ to the wall. Such a collision occurs once in a time interval $\Delta t = 2L_x/v_{ix}$, so the impulse per unit time, that is, the time average of the force on the wall, is

$$\bar{f}_i = \frac{2mv_{ix}}{2L_x/v_{ix}} = \frac{mv_{ix}^2}{L_x} . \tag{1.30}$$

Summation of this force over all of the molecules gives the force on the wall, and when divided by the area $S = L_yL_z$ it gives the pressure P:

$$\begin{aligned} P &= \sum_{i=1}^{N} \frac{\bar{f}_i}{S} = \sum_{i=1}^{N} \frac{mv_{ix}^2}{V} \\ &= \frac{1}{3V} \sum_{i=1}^{N} m(v_{ix}^2 + v_{iy}^2 + v_{iz}^2) = \frac{2}{3}\frac{1}{V} \sum_{i=1}^{N} \frac{1}{2} m\boldsymbol{v}_i^2 \\ &= \frac{2}{3}\frac{1}{V} E = \frac{N}{V\beta} . \end{aligned} \tag{1.31}$$

In the third equality in (1.31), we have used the fact that the motion of the gas must be isotropic, so that $\sum_i v_{ix}^2 = \sum_i v_{iy}^2 = \sum_i v_{iz}^2$. In the final equality, we have noticed the fact that the sum of the kinetic energies is the total energy E, and is equal to $3N/2\beta$ according to (1.28).

An ideal gas is known to satisfy the Boyle–Charles equation of state $PV = Nk_\mathrm{B}T$. Combining this with the present result gives $\beta = 1/k_\mathrm{B}T$, as mentioned before. However, you should remember that we have not defined the absolute temperature yet.

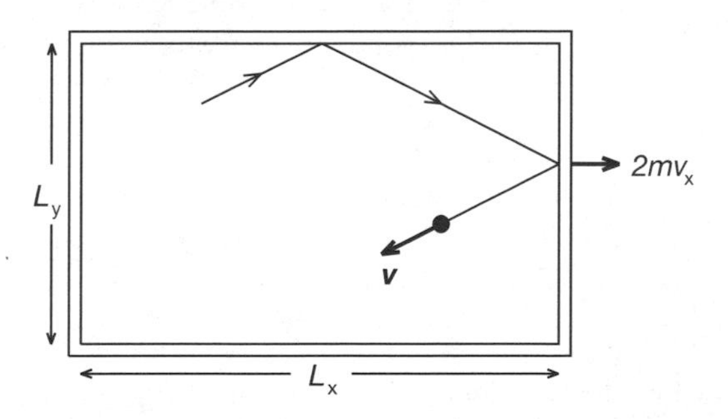

Fig. 1.5. Collisions of molecules with the walls are the origin of the pressure of a gas. In a collision with the right wall, a molecule gives an impulse $2mv_x$ to the wall

1.4 The Principle of Equal Probability

In this chapter, we have succeeded in reproducing the properties of an ideal gas in equilibrium. For the spatial distribution of the molecules, we assumed that each molecule can be anywhere in the container with equal probability, and obtained a uniform density. For the distribution in velocity space, we assumed similarly that each molecule can move with any velocity it likes, and obtained the Maxwell distribution, which has been confirmed experimentally. In both cases we assumed equal probabilities for microscopic motions that we cannot detect, except for imposing some obvious macroscopic constraints. Namely, for microscopic motions that we cannot observe or control, we allowed the molecules to do whatever they like. We then succeeded in obtaining the correct thermal properties. This success leads us to state a principle of equal probability, namely, "In thermal equilibrium under macroscopic constraints, every possible microscopic state allowed by the constraints is realized with equal probability." Here the macroscopic constraints are what are called state variables, such as the total energy or the volume of the system.

Strictly speaking, this principle is not exact. For example, in the derivation of the Maxwell distribution, molecules are allowed to move with speeds higher than the speed of light. However, in the Maxwell distribution obtained in this way, the probability of finding such high-speed molecules is so small that it is practically zero. Because there are a macroscopic number of molecules, the peak of the distribution becomes very sharp, and most of the microscopic realizations belong to states around the peak. Thus, even if the principle is not rigorously satisfied, the result obtained is essentially correct. In the following chapters, we construct the statistical physics of equilibrium states on the basis of this principle. The statistical physics thus constructed is then applied to several examples. From these examples, you will learn the paradoxical fact that as a result of every microscopic possibility being allowed, a unique thermal equilibrium is realized macroscopically. The origin of this paradox is the macroscopic number of molecules.

Exercise 1. Using the definition of a calorie and the fact that the mass of $1\,\mathrm{cm}^3$ of water is $1\,\mathrm{g}$, calculate the molar heat of water. Express the answer in units of the gas constant R.

Exercise 2. Iron has a molar heat of about $25\,\mathrm{J\,mol^{-1}\,K^{-1}}$. Calculate the heat capacity of a block of iron whose volume is $1\,\mathrm{m}^3$. Compare the result with that for the same volume of water. (The relative atomic mass of iron is 55.845 and its density is $7.86 \times 10^3\,\mathrm{kg\,m^{-3}}$.)

Exercise 3. Calculate the increase in the gravitational potential energy of $1\,\mathrm{mol}$ of water when it is brought from sea level to an altitude of $100\,\mathrm{m}$. Calculate the increase in the temperature of the water when this amount of work is supplied to the water to increase its internal energy instead of raising it in height.

Exercise 4. The internal energy of an ideal gas depends only on its temperature, and is independent of the volume V. Use this fact and the Boyle–Charles law to derive the relation between the constant-volume and constant-pressure heat capacities of an ideal gas, $C_{\mathrm{P}} = C_{\mathrm{V}} + Nk_{\mathrm{B}}$. (This relation is known as the Mayer relation.)

Exercise 5. The average velocity of a molecule of an ideal gas is given by $v = \sqrt{2k_{\mathrm{B}}T/m}$. Evaluate this velocity for an oxygen molecule in a gas at 300 K. (One mole of oxygen molecules weighs about 32 g.)

2

Entropy

In the previous chapter, we introduced the principle of equal probability. In this chapter we define entropy on the basis of this principle. Temperature is defined through this entropy. We develop a general theory here; in particular, we discuss the condition for thermal equilibrium under the constraint of a given total energy. It is shown that this is the condition that the entropy is maximized.

2.1 The Microcanonical Distribution

We consider a system enclosed by an adiabatic wall. By an adiabatic wall we mean that energy cannot be transferred through this wall. The system can be a gas, liquid, or solid. For this system, the volume V, the number of molecules N, and the total energy E are kept constant. These variables represent the only possible constraints for this system. Other than these constraints, we cannot place any restriction on the microscopic motions of the molecules. Therefore, according to the principle of equal probability, each molecule can do anything that is possible under these macroscopic constraints, and various microscopic states should be realized with equal probability in thermal equilibrium. Among these microscopic states, some states may be quite special, such that they will not be realized in reality. However, except for a few such exceptional states, almost all microscopic states will actually be realized as the state of the system changes temporally. Because of the vast number of possible microscopic states, the exceptional states will practically never be realized anyway; this is the conclusion of the previous chapter. We call this situation, where every possible microscopic state is realized with equal probability, the *microcanonical distribution.*

We write the total number of microscopic states allowed under the macroscopic constraints as $W(E, \delta E, V, N)$. Therefore, the probability of each microscopic state being realized is $1/W$. Here we have allowed some uncertainty δE in the total energy, and have counted microscopic states where the total energy is between E and $E + \delta E$. The reason for this will be clarified later. Using this W, we define the *entropy* $S(E, \delta E, V, N)$ as follows:

$$S(E, \delta E, V, N) = k_{\mathrm{B}} \ln W(E, \delta E, V, N) . \tag{2.1}$$

Furthermore, we define the *temperature* of the system as follows:

$$T = \left[\frac{\partial S(E, \delta E, V, N)}{\partial E} \right]^{-1} . \tag{2.2}$$

In this equation, the partial differentiation is done keeping δE, N, and V fixed. As we shall see later, the temperature thus defined coincides with the temperature defined by thermodynamics.

At this point you may have several questions.

- *Question 1*: The uncertainty in the energy δE looks arbitrary. If so, can the entropy and the temperature be defined without arbitrariness?
- *Question 2*: How can we count the number of microscopic states? In the case of a gas, the position and velocity of each molecule are continuous variables. Do two molecules with slightly different velocities belong to the same microscopic state or not?
- *Question 3*: Why is the uncertainty δE needed?

Before proceeding, let us answer these questions briefly.

First, we consider the effect of δE. We can show that the entropy and temperature do not depend on the choice of δE as long as δE is small enough. When δE is sufficiently small, W should be proportional to δE, i.e. $W = w \, \delta E$, where w is independent of δE. We can compare the value of the entropy for two different choices of δE, δE_1 and δE_2. The entropies S_1 and S_2 are calculated using the uncertainties δE_1 and δE_2, respectively:

$$S_1 = k_{\mathrm{B}} \ln (w \, \delta E_1) \tag{2.3}$$

and

$$S_2 = k_{\mathrm{B}} \ln (w \, \delta E_2) . \tag{2.4}$$

Now, the number of macroscopic states W is tremendously large when N is large, and the entropy S_i turns out to be proportional to N, as we shall see later, i.e. $S_i = k_{\mathrm{B}} O(N)$. Here, $O(N)$ means a number of the order of N. On the other hand, the difference between S_1 and S_2 turns out to be of the order of unity:

$$S_1 - S_2 = k_{\mathrm{B}} \ln \left(\frac{\delta E_1}{\delta E_2} \right) = k_{\mathrm{B}} O(1) . \tag{2.5}$$

Therefore, for macroscopic systems where N is of the order of 10^{22}, the difference between S_1 and S_2 is negligibly small compared with the value of S_i, and so the choice of δE is arbitrary. Because of this arbitrariness of δE, we shall not write δE in the argument of S in the following equations.

Second, we consider question 2. It is true that we cannot count the number of microscopic states if we consider the motion of molecules in classical mechanics. In classical mechanics, the state of a molecule is described by the continuous variables $\boldsymbol{r}$ and $\boldsymbol{v}$. Therefore, we cannot give a reasonable definition of the entropy. However, classical mechanics is not valid for the description of the motion of microscopic particles. At the microscopic level, the motion is governed by quantum mechanics. As we shall see in Chap. 4, the motion of molecules is quantized, and we can count the number of microscopic states even for a gas molecule in a container.

As a consequence of quantum mechanics, the energy is quantized also. That is, the kinetic energy of a molecule is no longer a continuous variable. This fact gives us an answer to question 3. Namely, we need a nonzero δE to have a reasonable behavior of W. If we make δE arbitrary small, W ceases to be a well-behaved function of E. It may happen for some energy E that there is no microscopically allowed state, but in the vicinity of that energy, W may become quite large. Finally, we remark on the derivative with respect to energy. Even if W is not a wild function of E for nonzero δE, it will change stepwise. However, each step is much smaller than the value of W there, and therefore it is possible to approximate W by a continuous function of E. The differentiation should be considered to be performed on this continuous function.

Example

For a better understanding of W and S, let us investigate a simple example of the counting of microscopic states. We consider here a situation where a large amount of money is distributed among a population of N people.[1] The total amount of money and the total number of people are macroscopic variables, which are fixed as constraints; that is, they are the state variables. If the money is in the form of a bar of gold, it can be divided almost continuously, and so it is hard to count the ways to distribute it among N people. This situation is similar to the situation where we treat a physical system by classical mechanics. However, in the case of money there are units, for example yen or cents, just as energy is quantized in quantum mechanics. If we use such units, the number of ways to distribute the money is countable.

The number of ways of distributing E yen to N people is given as follows:

$$W = \frac{(E+N-1)!}{E!(N!-1)} \simeq \left(\frac{E+N}{E}\right)^E \left(\frac{E+N}{N}\right)^N . \tag{2.6}$$

Here, to obtain the final result, Stirling's formula has been used. Thus the entropy and temperature are given by

$$S(E,N) = k_{\mathrm{B}} \ln W \simeq k_{\mathrm{B}} E \ln\left(1+\frac{N}{E}\right) + k_{\mathrm{B}} N \ln\left(1+\frac{E}{N}\right) \tag{2.7}$$

[1] This situation may seem unrelated to statistical physics, but the same equation appears later when we consider the heat capacity of a solid.

and

$$\frac{1}{T} = \frac{\partial S}{\partial E} = k_{\mathrm{B}} \ln \left(1 + \frac{N}{E}\right) . \tag{2.8}$$

Expressing E in terms of T, we obtain

$$E = \frac{N}{\exp(1/k_{\mathrm{B}}T) - 1} . \tag{2.9}$$

This form is known as the *Bose distribution function*, and we shall encounter it when we consider the vibration of two-atom molecules, black-body radiation, and the heat capacity of solids. A detailed discussion of this distribution function will be given in Chap. 10. You should notice that to obtain these results, we have implicitly taken δE to be 1 yen. Also, we have approximated the stepwise function $W(E, N)$ by a smooth function after using Stirling's formula, by treating E as a real number.

Remark

When two systems of the same size are united into one, E, N, and V are doubled. In this case the entropy should also be doubled. Otherwise, the temperature defined by (2.2) will change after the union. The entropy S (2.7) has this property. Similarly, the entropies of all the systems that we consider in this book have this property. If we want to define an entropy with this property starting from W, taking the logarithm of W is necessary. Then what is the reason why W was chosen as the starting point for obtaining S? I think that there is no a priori, logical reason to choose W. This choice was discovered by the genius Ludwig Boltzmann, probably after much trial and error. We can imagine how excited he was when he found that this entropy and the thermodynamic entropy were the same.

2.2 Number of States and Density of States

The entropy is defined by the number of microscopic states for a given energy. It can also be calculated from the *number of states* or from the *density of states*, both of which we define in this section. Any system has a state that has the lowest total energy out of all possible states. This is called the *ground state* in quantum mechanics. We take the origin of the energy so that the energy of the ground state is zero. We define the number of states $\Omega_0(E)$ as the number of microscopic states that have a total energy less than or equal to E. By definition, the number of states $\Omega_0(E)$ is an increasing function of E.

As in the case of the distribution of money considered above, the energy E may not be allowed to take a continuous set of values. However, since the total energy is the result of a summation over a macroscopic number of molecules, it should be tremendously larger than the unit of energy. Therefore it is legitimate to consider the number of states as a continuous function of E, and we can define the density of states as the derivative of the number of states with respect to E, i.e.

$$\Omega(E) = \frac{\mathrm{d}\Omega_0(E)}{\mathrm{d}E} \,. \tag{2.10}$$

The number of microscopic states W is nothing but $\Omega(E)$ multiplied by some energy uncertainty δE, i.e. $W = \Omega(E)\,\delta E$.

As seen in our example related to money, W increases exponentially with E. Therefore $\Omega(E)$ and $\Omega_0(E)$ also increase exponentially. Most macroscopic systems behave like this, and are called statistical-mechanically normal systems.[2] In this case we can use $\Omega_0(E)$ to calculate the entropy from $S = k_\mathrm{B} \ln[\Omega_0(E)]$. This is because of the following inequality, which can easily be shown to be satisfied from comparison of the areas shown in Fig. 2.1:

$$S = k_\mathrm{B} \ln[\Omega(E)\,\delta E] < k_\mathrm{B} \ln[\Omega_0(E)] < k_\mathrm{B} \ln[\Omega(E) E] \,. \tag{2.11}$$

If we calculate the difference between the right-hand and left-hand sides of these inequalities, we obtain

$$k_\mathrm{B}\{\ln[\Omega(E)E] - \ln[\Omega(E)\,\delta E]\} = k_\mathrm{B} \ln\left(\frac{E}{\delta E}\right) \,. \tag{2.12}$$

Since the entropy is a macroscopic quantity proportional to N, this difference is negligibly small compared with the entropy itself. Even if $E/\delta E$ is quite large, such as a billion or a trillion, the logarithm of it is not large.[3] This

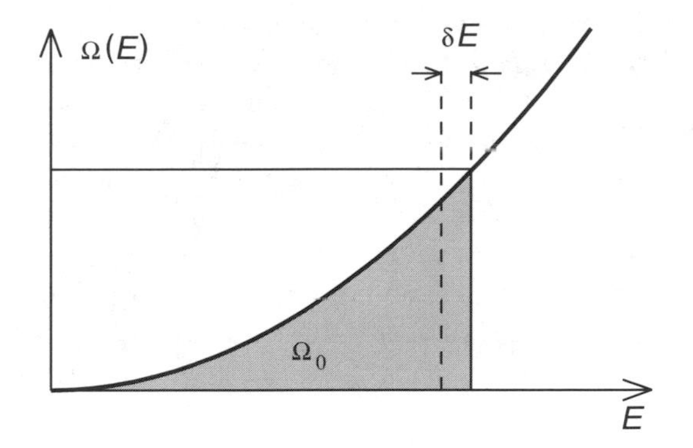

Fig. 2.1. Relationship between the energy and the number of states. The density of states $\Omega(E)$ is plotted. The number of states Ω_0 is given by the *shaded area below the curve.* It is larger than $\Omega\,\delta E$ but smaller than $\Omega \times E$

---

[2] Under special conditions it can happen that $\Omega(E)$ or W is not an increasing function of E. In the case of our example related to money, this happens when there is an upper limit on the amount of money that an individual can possess. A physical example is provided by a model where only a finite number of electronic states are taken into account.

[3] When $E/\delta E = 10^{30}$, $\ln(E/\delta E) = 71.4$. This value is negligible compared with $N \simeq 10^{24}$.

means that all the expressions in this inequality have effectively the same value, and any of these expressions can be used to calculate the entropy.

2.3 Conditions for Thermal Equilibrium

We consider two systems that are separated. We then couple these two systems with a weak interaction. After we have waited for a sufficiently long time, the two systems will reach a mutual equilibrium owing to the second law of thermodynamics. We seek a condition to describe this equilibrium state.

When the interaction is weak, the microscopic states of each system are almost the same as when the systems are isolated. As a matter of fact, the systems will interact through parts of their surfaces, and so the explicit effect of the interaction will be restricted to a region around those parts. Thus, we can assume that the density of states $\Omega(E)$ of each system will remain the same as in the isolated case. We distinguish the two systems by using subscripts I and II, and so we write the variables as E_{I}, N_{I}, and V_{I} for system I and as E_{II}, N_{II}, and V_{II} for system II.

2.3.1 Equilibrium Condition when only Energy is Exchanged

First we consider the case in which N_{I}, V_{I}, N_{II}, and V_{II} are kept constant, but energy can be transferred between the systems. In this case only the sum of the energies $E = E_{\mathrm{I}} + E_{\mathrm{II}}$ is conserved. We shall calculate the density of states of the total system $\Omega(E)$, or, more precisely, the number of microscopic states $\Omega(E)\,\delta E$. This is expressed in terms of the corresponding quantities for each system, Ω_{I} and Ω_{II}. When system I is in one of the microscopic states of energy E_{I}, system II can be in any of the microscopic states of energy E_{II} that satisfy $E - E_{\mathrm{I}} \leq E_{\mathrm{II}} \leq E - E_{\mathrm{I}} + \delta E$. The number of such states is $\mathcal{N}_{\mathrm{II}} = \Omega_{\mathrm{II}}(E - E_{\mathrm{I}})\,\delta E$. The number of microscopic states in the whole system in which system I has an energy between E_{I} and $E_{\mathrm{I}} + \mathrm{d}E_{\mathrm{I}}$ is then this number $\mathcal{N}_{\mathrm{II}}$ times the number of states in system I in this energy interval $\mathcal{N}_{\mathrm{I}} = \Omega_{\mathrm{I}}(E_{\mathrm{I}})\,\mathrm{d}E_{\mathrm{I}}$:

$$\mathcal{N}_{\mathrm{I}}\mathcal{N}_{\mathrm{II}} = \Omega_{\mathrm{I}}(E_{\mathrm{I}})\,\mathrm{d}E_{\mathrm{I}} \times \Omega_{\mathrm{II}}(E - E_{\mathrm{I}})\,\delta E\,. \tag{2.13}$$

The number of states of the whole system $\Omega(E)$ is obtained by summing this number over the possible range of E_{I}:

$$\Omega(E)\,\delta E = \int_0^E \mathrm{d}E_{\mathrm{I}}\,\Omega_{\mathrm{I}}(E_{\mathrm{I}}) \times \Omega_{\mathrm{II}}(E - E_{\mathrm{I}})\,\delta E\,. \tag{2.14}$$

According to the principle of equal probability, all these $\Omega(E)\,\delta E$ microscopic states are realized with equal probability in thermal equilibrium for a total energy E. Therefore, the probability $f(E_\mathrm{I})\,\mathrm{d}E_\mathrm{I}$ that system I has some energy between E_I and $E_\mathrm{I}+\mathrm{d}E_\mathrm{I}$ is $\mathcal{N}_\mathrm{I}\mathcal{N}_\mathrm{II}$ divided by $\Omega(E)\,\delta E$:

$$f(E_\mathrm{I})\,\mathrm{d}E_\mathrm{I} = \frac{\Omega_\mathrm{I}(E_\mathrm{I})\,\mathrm{d}E_\mathrm{I}\,\Omega_\mathrm{II}(E-E_\mathrm{I})\,\delta E}{\Omega(E)\,\delta E}\,. \tag{2.15}$$

Considering the fact that $\Omega_\mathrm{I}(E_\mathrm{I})$ and $\Omega_\mathrm{II}(E_\mathrm{II})$ are rapidly increasing functions of energy, we can expect that this product will show a very sharply peaked structure. In that case the system will almost always remain at around the position of the peak. That is, in equilibrium, the product $\Omega_\mathrm{I}(E_\mathrm{I})\Omega_\mathrm{II}(E-E_\mathrm{I})$ takes its maximum value. The way in which the peak becomes sharp for a macroscopic system is illustrated in Fig. 2.2.

The position of the peak is given by the maximum of $\ln[\Omega_\mathrm{I}(E_\mathrm{I})\Omega_\mathrm{II}(E-E_\mathrm{I})]$:

$$\begin{aligned}0 &= \frac{\mathrm{d}}{\mathrm{d}E_\mathrm{I}}k_\mathrm{B}\ln[\Omega_\mathrm{I}(E_\mathrm{I})\Omega_\mathrm{II}(E-E_\mathrm{I}] = \frac{\mathrm{d}}{\mathrm{d}E_\mathrm{I}}[S_\mathrm{I}(E_\mathrm{I})+S_\mathrm{II}(E-E_\mathrm{I})]\\ &= \frac{\partial S_\mathrm{I}(E_\mathrm{I})}{\partial E_\mathrm{I}} - \frac{\partial S_\mathrm{II}(E-E_\mathrm{I})}{\partial E}\,.\end{aligned} \tag{2.16}$$

That is, the total entropy $S(E) = S_\mathrm{I}(E_\mathrm{I}) + S_\mathrm{II}(E_\mathrm{II})$ is maximized in thermal equilibrium under the constraint that the total energy is fixed. Now we recall that the temperature is defined by the derivative of the entropy with respect to the energy, i.e. $\partial S/\partial E = 1/T$. Therefore, (2.16) indicates that the temperatures of the two systems become the same when they are brought to equilibrium through exchange of energy.

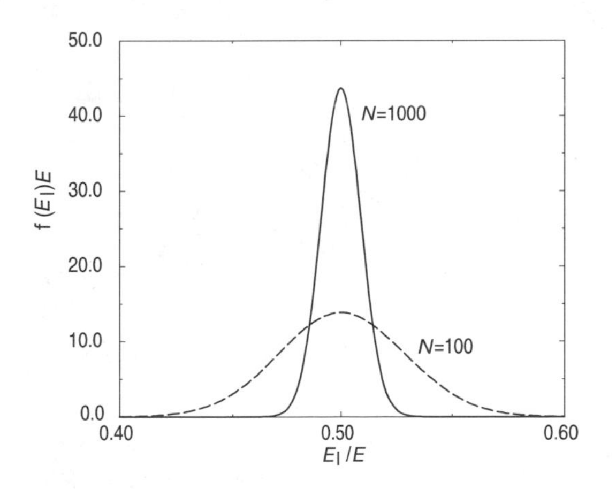

Fig. 2.2. The probability that system I has energy E_I, $f(E_\mathrm{I}) = \Omega_\mathrm{I}(E_\mathrm{I})\Omega_\mathrm{II}(E-E_\mathrm{I})/\Omega(E)$. To normalize the figure, what is actually plotted is $f(E_\mathrm{I})E$ as a function of E_I/E. This figure shows the probability when the two systems are ideal gases with equal numbers N of molecules. In this case $\Omega_\mathrm{I}(E_\mathrm{I}) \propto E_\mathrm{I}^{3N/2}$. The *dashed line* and *solid line* show the probability when $N = 100$ and $N = 1000$, respectively. The *peak* narrows in proportion to $\sqrt{N}$, so for macroscopic systems the *peak* is very narrow, like that shown in Fig. 1.2

2.3.2 Equilibrium Condition when Molecules are Exchanged

Next we consider the case in which two systems are connected by a small hole, and molecules can be exchanged between them. Since the molecules carry energy, energy is also exchanged. The energy and the number of molecules in each system are not conserved, but the total number of molecules $N = N_{\mathrm{I}} + N_{\mathrm{II}}$ and the total energy $E = E_{\mathrm{I}} + E_{\mathrm{II}}$ are conserved. We already know the density of states of the total system when each of the two systems has a fixed number of molecules. The density of states in the present case is obtained by summing the density of states for a fixed number over the possible distributions of the molecules, i.e. for $0 \leq N_{\mathrm{I}} \leq N$. Thus, the density of states of the total system is

$$\Omega(E, N) = \sum_{N_{\mathrm{I}}=0}^{N} \int_0^E \Omega_{\mathrm{I}}(E_{\mathrm{I}}, N_{\mathrm{I}}) \Omega_{\mathrm{II}}(E - E_{\mathrm{I}}, N - N_{\mathrm{I}}) \, \mathrm{d}E_{\mathrm{I}} \, . \tag{2.17}$$

The probability for system I to have an energy between E_{I} and $E_{\mathrm{I}} + \mathrm{d}E_{\mathrm{I}}$ and N_{I} molecules is

$$f(E_{\mathrm{I}}, N_{\mathrm{I}}) \, \mathrm{d}E_{\mathrm{I}} = \frac{\Omega_{\mathrm{I}}(E_{\mathrm{I}}, N_{\mathrm{I}}) \Omega_{\mathrm{II}}(E - E_{\mathrm{I}}, N - N_{\mathrm{I}}) \, \mathrm{d}E_{\mathrm{I}}}{\Omega(E, N)} \, . \tag{2.18}$$

Thermal equilibrium is attained when this probability reaches its maximum. Since the logarithm of the numerator is the entropy of the total system $S(E, N) = S_{\mathrm{I}}(E_{\mathrm{I}}, N_{\mathrm{I}}) + S_{\mathrm{II}}(E_{\mathrm{II}}, N_{\mathrm{II}})$, the condition is that the total entropy is maximized under the constraint of fixed E and N:

$$\begin{aligned} \left(\frac{\partial S(E, N)}{\partial E_{\mathrm{I}}} \right)_{E, N, N_{\mathrm{I}}} &= \left(\frac{\partial S_{\mathrm{I}}(E_{\mathrm{I}}, N_{\mathrm{I}})}{\partial E_{\mathrm{I}}} \right)_{N_{\mathrm{I}}} - \left(\frac{\partial S_{\mathrm{II}}(E - E_{\mathrm{I}}, N - N_{\mathrm{I}})}{\partial E} \right)_{N_{\mathrm{I}}} \\ &= 0 \end{aligned} \tag{2.19}$$

and

$$\begin{aligned} \left(\frac{\partial S(E, N)}{\partial N_{\mathrm{I}}} \right)_{E, N, E_{\mathrm{I}}} &= \left(\frac{\partial S_{\mathrm{I}}(E_{\mathrm{I}}, N_{\mathrm{I}})}{\partial N_{\mathrm{I}}} \right)_{E_{\mathrm{I}}} - \left(\frac{\partial S_{\mathrm{II}}(E - E_{\mathrm{I}}, N - N_{\mathrm{I}})}{\partial N} \right)_{E_{\mathrm{I}}} \\ &= 0 \, . \end{aligned} \tag{2.20}$$

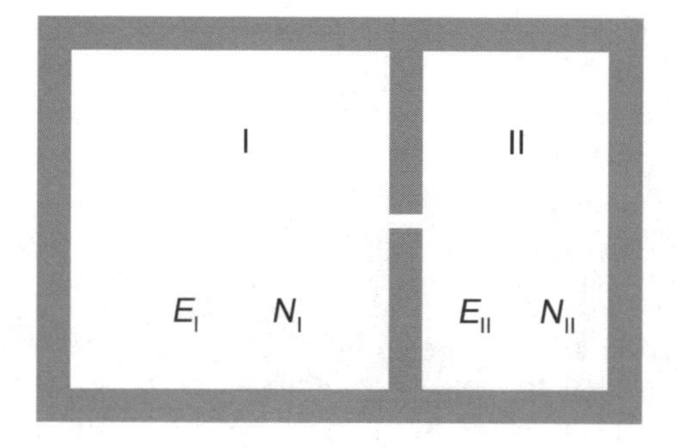

Fig. 2.3. Two systems connected by a small hole

We define the *chemical potential* μ by the following equation:

$$\left(\frac{\partial S}{\partial N}\right)_{E,V} = -\frac{\mu}{T}. \tag{2.21}$$

The conditions for thermal equilibrium are $T_{\rm I} = T_{\rm II}$ and $\mu_{\rm I} = \mu_{\rm II}$; that is, the temperature and the chemical potential must both have equal values in the two systems.

2.3.3 Equilibrium Condition when Two Systems Share a Common Volume

This time we consider two systems that are connected by a movable wall, for example two systems in a cylinder separated by a piston as shown in Fig. 2.4. The motion of the wall transmits energy between the systems, but the numbers of molecules are conserved in both systems. In this case we count the number of microscopic states in system I under the condition that the system has an energy between $E_{\rm I}$ and $E_{\rm I} + {\rm d}E_{\rm I}$ and a volume between $V_{\rm I}$ and $V_{\rm I} + {\rm d}V_{\rm I}$. We write the result as $\Omega_{\rm I}(E_{\rm I}, V_{\rm I})\,{\rm d}E_{\rm I}\,{\rm d}V_{\rm I}$. The density of states of the total system is then

$$\Omega(E,V) = \int_0^E {\rm d}E_{\rm I} \int_0^V {\rm d}V_{\rm I}\, \Omega_{\rm I}(E_{\rm I}, V_{\rm I})\Omega_{\rm II}(E - E_{\rm I}, V - V_{\rm I}), \tag{2.22}$$

and the probability of system I having an energy between $E_{\rm I}$ and $E_{\rm I} + {\rm d}E_{\rm I}$ and a volume between $V_{\rm I}$ and $V_{\rm I} + {\rm d}V_{\rm I}$ is

$$f(E_{\rm I}, V_{\rm I})\,{\rm d}E_{\rm I}\,{\rm d}V_{\rm I} = \frac{\Omega_{\rm I}(E_{\rm I}, V_{\rm I})\Omega_{\rm II}(E - E_{\rm I}, V - V_{\rm I})\,{\rm d}E_{\rm I}\,{\rm d}V_{\rm I}}{\Omega(E,V)}. \tag{2.23}$$

The condition for thermal equilibrium is that this probability has its maximum value; this is the condition that the total entropy $S_{\rm I}(E_{\rm I}, V_{\rm I}) + S_{\rm II}(E - E_{\rm I}, V - V_{\rm I})$ is maximized with respect to $E_{\rm I}$ and $V_{\rm I}$. We define the pressure as follows:

$$\left(\frac{\partial S}{\partial V}\right)_{E} = \frac{P}{T}. \tag{2.24}$$

The equilibrium condition is that the pressures of the two systems are equal, together with the temperatures, namely $T_{\rm I} = T_{\rm II}$ and $P_{\rm I} = P_{\rm II}$.

As we have seen, the total entropy is maximized in thermal equilibrium. We have defined the entropy, and have defined the temperature, pressure, and chemical potential as derivatives of the entropy. We have then rewritten the condition for thermal equilibrium in terms of those variables. Are these variables the same as those which occur in thermodynamics or mechanics, and are measured in experiments? This is a natural question that you should have. The answer to this question is given in Chap. 4, where we discuss an ideal gas by using statistical physics. There we shall find that the statistical-mechanical definitions of these variables coincide with the traditional definitions.

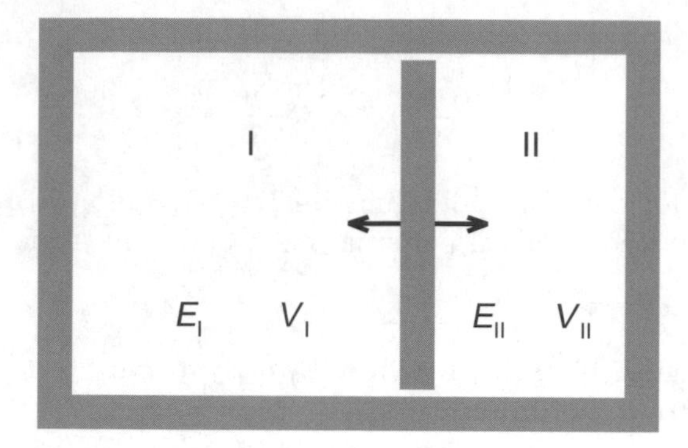

Fig. 2.4. Two systems separated by a movable wall

2.4 Thermal Nonequilibrium and Irreversible Processes

We have seen that the entropy is maximized in a state of thermal equilibrium. Here we consider the entropy in a state of thermal *nonequilibrium*. Thermal-nonequilibrium states are caused by actions from outside the system. For example, the earth as a whole is not in thermal equilibrium. There is a nearly constant input of energy from the sun in the form of electromagnetic radiation. This energy is radiated, in turn, from the earth to the cosmos. In the presence of this flow of energy, the earth cannot be in an equilibrium state, and various meteorological phenomena occur and life is supported. Another example is a conductor connected to a battery. In this case an electric current flows in the conductor. The systems in these examples are in quasi-stationary nonequilibrium states. That is, energy is continuously put into the system, and the nonequilibrium state lasts for a long time. On the other hand, there are also situations where a nonequilibrium state is prepared by some means or other, and after that the system is left to evolve by itself without further input of energy from outside. One example is that of two systems at different temperatures, placed in thermal contact at some time. Such a nonequilibrium state, when left alone, approaches a thermal-equilibrium state as time elapses. This process is irreversible. A thermal-nonequilibrium state changes into an equilibrium state spontaneously, but not vice versa. We shall argue that in such an irreversible process the entropy always increases.

For that purpose, we note that a thermal-nonequilibrium state can be created by imposing various constraints on the system. One example of a typical nonequilibrium state is a system in which the temperature depends on position. Another example is that of a cup of water to which a droplet of ink has just been added. In order to keep these systems in their initial condition, we would need to divide the system into many small cells with adiabatic barriers between them in the case of the first example, and to wrap the droplet of ink in the case of the second example. A thermal-equilibrium state would be approached after these constraints were removed. The molecules acquire more freedom when the constraints are removed. Thus, it is evident that the number of microscopic states increases when this removal happens. Therefore, we can conclude that the entropy, defined as the logarithm of the number of microscopic states, is larger in the thermal-equilibrium state. That is, as the system evolves from a nonequilibrium state to an equilibrium state, the entropy increases.

In other words, the microscopic states allowed in a nonequilibrium state are a subset of the microscopic states allowed in the equilibrium state. The principle of equal probability tells us that any microscopic state in the nonequilibrium state is realized in the equilibrium state with some probability, in principle. However, when the constraints that are needed to realize the nonequilibrium state are removed, the number of microscopic states allowed increases so tremendously that the probability that the system will return to a state in the original subset is vanishingly small.

Exercise 6. Consider the problem of distributing E yen to a population of N people, where, in this case, $E < N$ and everyone is allowed to possess at most 1 yen. Calculate the entropy and temperature of this model. Express E as a function of T.

3

The Partition Function and the Free Energy

In the previous chapter, we discussed a system under the constraint that the total energy is conserved. However, this condition is difficult to achieve in a real experimental situation. Therefore, in this chapter, we discuss a system for which energy is not conserved. Instead of a system for which energy is conserved, we consider a system in thermal equilibrium with a heat bath. In this situation, the free energy plays an important role. We derive an expression for the free energy. This chapter concludes our discussion of the basics of the statistical physics of equilibrium systems.

3.1 A System in a Heat Bath

It is difficult to isolate a system energetically from its environment. Energy may enter the system by thermal conduction, electromagnetic radiation (e.g. light), or sound, for example. Therefore, it is more appropriate to allow the energy of the system to fluctuate, and to discuss the probability of the system having various values of its energy. In this case the state of the system depends on its environment. Therefore, we must first prepare a well-defined environment. In addition to the system under consideration, we prepare a system that is itself in thermal equilibrium, and has a much larger heat capacity than the system under investigation. This larger system is called a *heat bath*, and we surround the system under consideration with it. The heat bath forms a well-defined environment. Since the heat capacity of the heat bath is much larger than that of the system under consideration, the amount of energy exchanged between the systems is negligibly small compared with the total energy of the heat bath. Therefore, the heat bath can be considered to be at a fixed temperature. We are going to examine the probability distribution of the energy of a system surrounded by a heat bath at temperature T.

3.1.1 Canonical Distribution

In this chapter, we consider a situation in which only energy is exchanged between the system and the heat bath. Other situations, in which volume or molecules are exchanged, will be treated in Chaps. 8 and 10, respectively.

In the previous chapter we discussed the condition for the thermal equilibrium of two systems when energy is exchanged. To utilize that discussion, we shall call the system under investigation system I, and the heat bath system II. In Chap. 2, we found the most probable distribution of energy between the two systems. Here, we discuss how the energy fluctuates around this most probable distribution. In this case, system I need not be a macroscopic system; it may be a microscopic system such as a single molecule. If system I is macroscopic, the most probable energy is practically always realized; the temperature of the system is defined and has the same value as that of the heat bath. It should be noted, however, that a temperature cannot be defined for system I if it is a microscopic system, and in that case the distribution of the energy can be rather wide, as we shall see later.

First, we consider the probability $f(E_{\mathrm{I}})$ that system I is in a microscopic state of energy E_{I}. The total system, i.e. system I plus system II, can be considered by use of the microcanonical distribution for a total energy E_{t}. The probability $f(E_{\mathrm{I}})$ is proportional to the number of microscopic states in which system I is in the given state. Since system I is already in a given microscopic state, this number is given by the number of microscopic states of system II at energy $E_{\mathrm{II}} = E_{\mathrm{t}} - E_{\mathrm{I}}$. Therefore,

$$f(E_{\mathrm{I}}) \propto \Omega_{\mathrm{II}}(E_{\mathrm{t}} - E_{\mathrm{I}}) \propto \frac{\Omega_{\mathrm{II}}(E_{\mathrm{t}} - E_{\mathrm{I}})}{\Omega_{\mathrm{II}}(E_{\mathrm{t}})} . \tag{3.1}$$

Here Ω_{II} is the density of states of the heat bath (system II), and the denominator in the final form is simply a constant that is introduced for later convenience. Because system II is much larger than system I, the condition $E_{\mathrm{t}} \gg E_{\mathrm{I}}$ should be satisfied.[1] We can expand the right-hand side with respect to E_{I}:

$$\begin{aligned} f(E_{\mathrm{I}}) &\propto \frac{\Omega_{\mathrm{II}}(E_{\mathrm{t}} - E_{\mathrm{I}})}{\Omega_{\mathrm{II}}(E_{\mathrm{t}})} \\ &= \exp\left[\frac{1}{k_{\mathrm{B}}}\left\{S_{\mathrm{II}}(E_{\mathrm{t}} - E_{\mathrm{I}}) - S_{\mathrm{II}}(E_{\mathrm{t}})\right\}\right] \\ &\simeq \exp\left[-\frac{E_{\mathrm{I}}}{k_{\mathrm{B}}}\frac{\partial S_{\mathrm{II}}(E_{\mathrm{t}})}{\partial E}\right] \\ &= \exp\left[-\frac{E_{\mathrm{I}}}{k_{\mathrm{B}}T}\right] . \end{aligned} \tag{3.2}$$

[1] This is the condition that system II can behave as a heat bath.

This gives the probability that system I is in a microscopic state with energy E_{I}. This is called the *canonical distribution.*

3.1.2 Application to a Molecule in Gas

In Chap. 1, we derived the Maxwell distribution, which describes the probability of a gas molecule having a particular velocity. If we consider one molecule in a gas as system I and the rest of the gas as a heat bath, then the Maxwell distribution (1.29) can be derived from the canonical distribution of the kinetic energy of the molecule. Since system I is microscopic, the energy distribution is quite broad.

When the gas is in a gravitational field, the energy of a molecule consists of its kinetic energy and potential energy:

$$\varepsilon = \frac{mv^2}{2} + mgh \,, \tag{3.3}$$

where h is the altitude and g is the acceleration due to gravity. Therefore, the probability $f(v,h)$ of finding a molecule at altitude h with speed v in a gas at temperature T is

$$f(v,h) \propto \exp\left[-\left(\frac{mv^2}{2k_{\mathrm{B}}T} + \frac{mgh}{k_{\mathrm{B}}T}\right)\right] . \tag{3.4}$$

We can estimate from this equation how the density of oxygen in the air decreases as we climb a mountain. The mass m of an oxygen molecule is 5.33×10^{-26} kg. Therefore, the ratio of the density of oxygen molecules at $h = 3000$ m to that at $h' = 0$ m at $T = 300$ K is

$$\exp\left(-\frac{mg(h-h')}{k_{\mathrm{B}}T}\right) = \mathrm{e}^{-0.378} = 0.68 \,. \tag{3.5}$$

Because of this decrease in the density of oxygen, you need oxygen cylinders to climb Mount Everest.

You might think that this equation could be used to determine k_{B}. Namely, if we measure how the density or pressure of the air decreases with altitude, we can determine m/k_{B}, and then, if we know m, k_{B} can be determined. However, this is not a good idea, because the ratio m/k_{B} is already known from macroscopic measurements. If m and k_{B} are multiplied by the Avogadro constant N_{A}, $mN_{\mathrm{A}} \simeq 32$ g is the molar mass of oxygen, and $k_{\mathrm{B}}N_{\mathrm{A}} = R$ is the gas constant, known from the equation of state of an ideal gas. Therefore, to determine the value of k_{B} we need to use something else, whose mass is known independently.

In 1908, Perrin measured the vertical distribution of resin particles in water and succeeded in determining the value of k_{B} independently of the

Avogadro number. He also showed that the values of these quantities were consistent. To prepare many resin particles of the same size, Perrin dissolved a resin in alcohol. The solution was then mixed with water to precipitate out the resin as small particles. Since the particles had various sizes, a centrifuge was used to separate the particles according to size. One kilogram of resin was processed over several months, and Perrin obtained only a tiny amount of resin particles with sizes that were sufficiently well matched. The size and density of the particles were measured, and the particles were then suspended in water. The vertical distribution was measured by counting particles in the field of view of a microscope with a depth of focus of $2\,\mu\mathrm{m}$; that is, particles that were in focus, which floated at heights between h and $h + 2\,\mu\mathrm{m}$, were counted. The dependence of the resulting number on h was plotted, and fitted to a dependence on h of the form (3.5). Since the mass m of a particle could be determined from the mass density of the resin and the diameter of the particles, this fitting process gave the value of k_B. This experiment to determine k_B brought Perrin a Nobel Prize.

3.2 Partition Function

The probability of the canonical distribution (3.2) is not normalized. Let us now normalize it. We number the microscopic states of system I, and write the energy of the ith state as E_i. It should be noted that it is possible that different microscopic states may have the same energy E_i. We first calculate the following sum Z over all microscopic states:

$$Z(T, V, N) = \sum_i \exp\left(-\frac{E_i}{k_\mathrm{B}T}\right) . \tag{3.6}$$

The normalized probability is then

$$f\left(E_i\right) = \frac{\mathrm{e}^{-E_i/k_\mathrm{B}T}}{Z} . \tag{3.7}$$

The normalization factor Z introduced here is called the *partition function*. It is a function of the temperature, the volume, and the number of molecules, and plays an important role, as we shall see in the next section. However, before discussing its importance, we shall rewrite the above equation in a more convenient form. For a macroscopic system, there are infinitely many microscopic states. Therefore, it is impossible to carry out the summation in (3.6). We need to calculate the sum by replacing the summation with an integral. There are many microscopic states between energies E and $E + \mathrm{d}E$. The number of states in this range is given by $\Omega_\mathrm{I}(E)\,\mathrm{d}E$, where $\Omega_\mathrm{I}(E)$ is the density of states. On the other hand, for all the states with energy E_i in this range, we can write $\exp(-E_i/k_\mathrm{B}T)$ as $\exp(-E/k_\mathrm{B}T)$. The summation over the states between E and $E + \mathrm{d}E$ can then be written as

$$\sum_{E \le E_i < E+\mathrm{d}E} \exp\left(-\frac{E_i}{k_\mathrm{B}T}\right) = \exp\left(-\frac{E}{k_\mathrm{B}T}\right) \sum_{E \le E_i < E+\mathrm{d}E} 1$$
$$= \exp\left(-\frac{E}{k_\mathrm{B}T}\right) \Omega_\mathrm{I}(E)\,\mathrm{d}E\,. \qquad (3.8)$$

Summing over all the whole range of energies, we obtain Z in the form of an integral:

$$Z = \int_0^\infty \mathrm{e}^{-E/k_\mathrm{B}T} \Omega_\mathrm{I}(E)\,\mathrm{d}E\,. \qquad (3.9)$$

3.3 Free Energy

We now introduce the *free energy* $F(T, V, N)$, derived from the partition function $Z(T, V, N)$:

$$F(T, V, N) \equiv -k_\mathrm{B}T \ln Z(T, V, N)\,. \qquad (3.10)$$

We shall show that this function is equal to $U - ST$ for a macroscopic system, where U is the internal energy of the system. Therefore, F coincides with the Helmholtz free energy of thermodynamics, and all possible information about the equilibrium state is known once this function has been obtained. To show this equality, we shall evaluate Z. We rewrite Z using the entropy $S = k_\mathrm{B} \ln(\Omega_\mathrm{I}\,\delta E)$:

$$Z(T, V, N) = \int_0^\infty \mathrm{d}E\, \mathrm{e}^{-E/k_\mathrm{B}T} \Omega_\mathrm{I}(E)$$
$$= \int_0^\infty \mathrm{d}E \exp\left[-\frac{1}{k_\mathrm{B}T}\{E - S(E, V, N)T\}\right] \frac{1}{\delta E}\,. \qquad (3.11)$$

In this integral, E and S are both macroscopic variables proportional to N. Thus, the integrand changes drastically around the minimum of $E - ST$, which occurs at E^*. Since the integrand becomes exponentially small once E deviates from E^*, the integral can be evaluated from the integrand in the vicinity of E^*. The position of the minimum E^* is determined by the following equation:

$$0 = \frac{\partial}{\partial E}(E - ST) = 1 - \left(\frac{\partial S}{\partial E}\right)_{T,V,N} T\,. \qquad (3.12)$$

It should be noted that the temperature in this equation is that of the heat bath. When we discussed the microcanonical distribution, the temperature of the system was given by

$$\left(\frac{\partial S}{\partial E}\right)_{V,N} = \frac{1}{T}\,. \qquad (3.13)$$

Here, a similar equation determines E^*.

The integrand is the probability that energy of the heat bath, of temperature T, is E. So E^* is the most probable energy of the system, and the sharp peak means that the fluctuation in the energy is small for a macroscopic system. Thus E^* is also the average energy $\langle E \rangle$ of the system. The sharpness of the peak is illustrated in Fig. 3.1.

To perform the integration, we expand $E - ST$ around E^*:

$$\begin{aligned} E - ST &= E^* - S(E^*, V, N)T + (E - E^*) \\ &\quad - (E - E^*)\frac{\partial S}{\partial E}T - \frac{1}{2}(E - E^*)^2 \frac{\partial^2 S}{\partial E^2}T + \cdots \\ &\simeq E^* - S(E^*, V, N)T + \frac{1}{2}(E - E^*)^2 \frac{1}{TC} \\ &\quad + O[(E - E^*)^3]. \end{aligned} \tag{3.14}$$

Here (3.12), i.e. $(\partial S/\partial E)T = 1$, has been used, and we have put

$$\frac{\partial^2 S}{\partial E^2} = -\frac{1}{T^2 C}. \tag{3.15}$$

C is a positive constant, which is actually the heat capacity of the system. The integral reduces to a Gaussian integral in this approximation, and the result is

$$Z = \left(2\pi k_{\mathrm{B}} T^2 C\right)^{1/2} \frac{1}{\delta E} \exp\left[-\frac{E^* - S(E^*, V, N)T}{k_{\mathrm{B}} T}\right]. \tag{3.16}$$

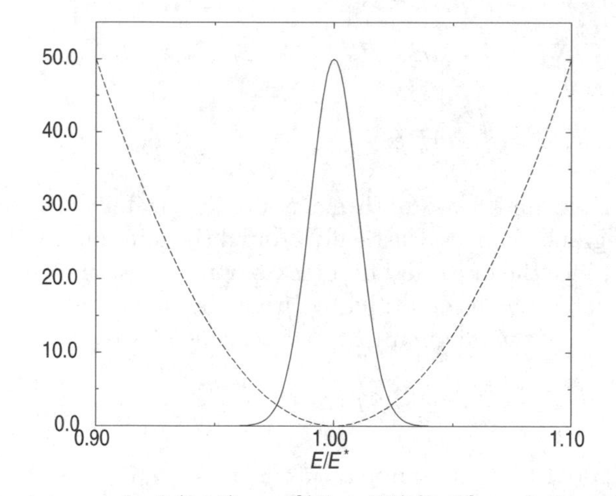

Fig. 3.1. The integrand of (3.11), $\exp[(E - ST)/k_{\mathrm{B}}T]$, and the argument of the exponential, $(E - ST)/k_{\mathrm{B}}T$. The *dashed line* shows the behavior of $(E - ST)/k_{\mathrm{B}}T$ around the minimum for a system consisting of a sample of an ideal gas with 10 000 molecules. The *solid line* shows $\exp[-(E-ST)/k_{\mathrm{B}}T]$. Since $(E-ST)/k_{\mathrm{B}}T$ increases in proportion to N, the number of molecules, the width of the integrand decreases in inverse proportion to $\sqrt{N}$, and it becomes very narrow for a macroscopic system

Hence,

$$F = -k_{\mathrm{B}}T \ln Z$$
$$= [E^* - S(E^*, V, N)T] - k_{\mathrm{B}}T \ln\left[\frac{\left(2\pi k_{\mathrm{B}}T^2 C\right)^{1/2}}{\delta E}\right]. \quad (3.17)$$

In this equation, the first term is a macroscopic quantity proportional to N. Compared with this term, the second term is negligibly small: even if the argument of the logarithm is macroscopic, the logarithm is of the order of one. As a result, we obtain

$$F = E^* - ST. \quad (3.18)$$

Namely, while F is equal to $-k_{\mathrm{B}}T \ln Z$, it is also the minimum value of $E - ST$ in the argument of the exponential in the integrand of (3.11), which gives Z.

The fact that the minimum of $E - ST$ gives the most probable distribution, namely thermal equilibrium, under the constraint of a given temperature and volume, is important. Here this function is minimized with respect to the energy E. However, we can extend this rule. It sometimes happens that a system is characterized by an additional state variable, say X, in addition to T, V, and N. An example of X might be the magnetization in the case of a ferromagnetic system. In this case, the summation over the microscopic states in the calculation of Z can be done under the constraint of a given value of X, and Z and F are given by functions of T, V, N, and X. The value of X in thermal equilibrium is given by the value at which F is a minimum. This is because the partition function is a maximum for that value of X, which means that the number of microscopic states is at its largest, and so one of them is almost always realized owing to the principle of equal probability. In Chap. 7, we apply this principle to discuss a phase transition in a magnetic material. This principle is also used in Landau's theory of second-order phase transitions, discussed in Chap. 9.

3.4 Internal Energy

In thermodynamics, the *internal energy* is introduced, and is usually written as U. It is the energy of a system after any mechanical energy associated with the center of mass of the system has been removed. Since it is usually trivial to remove such mechanical energy, we shall assume that "energy" in this section means the energy without any such external energy.

In the microcanonical distribution, the energy E is given and the temperature is determined as a function of E. There is no fluctuation in the energy, and the internal energy U is nothing but this energy E. In the canonical distribution, on the other hand, the temperature is given, and the microscopic states, with energy E_i, are distributed according to the probability (3.7). In this case U is given by $\langle E_i \rangle$. Since the distribution function of the energy is sharply peaked, the average coincides with the most probable value of the energy, E^*. Therefore, the relationship $F = U - ST$ can be established.

Let us now give a useful equation for U, and show that $U = \langle E \rangle$ actually coincides with E^*. From the definition of a thermal average in equilibrium, U can be expressed and rewritten as

$$U = \sum_i \frac{E_i \mathrm{e}^{-E_i/k_\mathrm{B}T}}{Z} = -\frac{\partial}{\partial \beta} \ln Z \,, \tag{3.19}$$

where $\beta = 1/k_\mathrm{B}T$ as usual. That is, U is obtained from the derivative of $\ln Z$. Using F for $\ln Z$, we can show that this average coincides with the most probable energy E^*:

$$\begin{aligned} U &= \frac{\partial}{\partial \beta}\left(\frac{F}{k_\mathrm{B}T}\right) \\ &= \frac{1}{k_\mathrm{B}T}\frac{\partial F}{\partial \beta} + F\frac{\partial}{\partial \beta}\left(\frac{1}{k_\mathrm{B}T}\right) \\ &= -T\frac{\partial F}{\partial T} + F = TS + F = E^* \,. \end{aligned} \tag{3.20}$$

In the last line we have used a well-known relation from thermodynamics, $(\partial F/\partial T)_V = -S$. We shall derive this relation in the next section.

3.5 Thermodynamic Functions and Legendre Transformations

The entropy that we defined in Chap. 2 is a function of the energy E, the volume V, and the number of molecules N, and its total differential is

$$\mathrm{d}S = \frac{1}{T}\mathrm{d}E + \frac{P}{T}\mathrm{d}V - \frac{\mu}{T}\mathrm{d}N \,. \tag{3.21}$$

This equation is satisfied in thermal equilibrium. Thus we can use U instead of E and rewrite the total differential of the entropy as a total differential of U:

$$\mathrm{d}U = T\,\mathrm{d}S - P\,\mathrm{d}V + \mu\,\mathrm{d}N \,. \tag{3.22}$$

This equation means that we should consider the internal energy U as a function of S, V, and N.

On the other hand, the free energy F defined in the previous section is a function of T, V, and N by definition. We can obtain the total differential of F using $F = U - ST$:

$$\mathrm{d}F = \mathrm{d}U - \mathrm{d}(TS) = \mathrm{d}U - S\,\mathrm{d}T - T\,\mathrm{d}S = -S\,\mathrm{d}T - P\,\mathrm{d}V + \mu\,\mathrm{d}N \,. \tag{3.23}$$

From this equation, we can obtain $(\partial F/\partial T)_{V,N} = -S$, for example, which was used in the previous section. In this way, we can change the independent variables from the set (S, V, N) to the set (T, V, N) by subtracting ST from U. This operation is called a *Legendre transformation.*

A further Legendre transformation defines the Gibbs free energy G as $G = F + PV$. This free energy is a function of (T, P, N):

$$dG = -S\,dT + V\,dP + \mu\,dN\,. \tag{3.24}$$

This Gibbs free energy has an interesting property. Among T, P, and N, only N is an extensive variable. Therefore, the following relation should be satisfied:

$$G(T, P, \alpha N) = \alpha G(T, P, N)\,. \tag{3.25}$$

This equation means that $G(T, P, N) = g(T, P)N$. Here g is the Gibbs free energy per molecule. On the other hand, from the derivative of G, we know that $(\partial G/\partial N)_{T,P} = \mu$. Thus we can conclude that $g(T, P) = \mu$, i.e.

$$G(T, P, N) = \mu(T, P)N\,. \tag{3.26}$$

U, F, and G are connected by a Legendre transformation and collectively called *thermodynamic functions.* They have different sets of independent variables. The variables associated with each function are called its *natural variables.* It is important to realize that this association is not arbitrary: once a thermodynamic function has been given as a function of its natural variables, we can obtain the values of any state variables in thermal equilibrium, but this is not so if the function is given in terms of another set of variables. For example, once we have obtained $F(T, V, N)$, S, P, and μ can be derived by differentiating F with respect to T, V, and N, respectively. Furthermore, other thermodynamic functions can be obtained by Legendre transformations. On the other hand, suppose that the internal energy U has been obtained as a function of (T, V, N); of these variables, T is not a natural variable. In this case we cannot derive S from an expression for $U(T, V, N)$. In fact, it is known, and is shown in the next chapter, that $U(T, V, N) = (3/2)Nk_{\mathrm{B}}T$ for an ideal gas composed of monatomic molecules. We cannot derive S or P from this function.

3.6 Maxwell Relations

From the fact that state variables are given by derivatives of thermodynamic functions, we can obtain important relations between their derivatives. Consider F as an example. Since it is differentiable, its second-order derivative can be calculated in two ways and the results should be the same. That is, the following equation is satisfied:

$$\frac{\partial^2}{\partial V \partial T} F = \frac{\partial^2}{\partial T \partial V} F\,. \tag{3.27}$$

Putting $(\partial F/\partial T)_V = -S$ and $(\partial F/\partial V)_T = -P$ into the two sides of this equation, we obtain the equation

$$\left(\frac{\partial S}{\partial V}\right)_T = \left(\frac{\partial P}{\partial T}\right)_V . \tag{3.28}$$

This kind of relation can be derived using other thermodynamic functions as well. All such relations are called *Maxwell relations.*

One application of these relations is to the determination of entropy. Entropy cannot be measured by simply reading a gauge on an apparatus, as can be done for temperature or pressure. Instead, (3.28) can be used to find out how S changes when the volume changes at a given temperature. Similarly, from an equation obtained from the derivative of G,

$$\left(\frac{\partial S}{\partial P}\right)_T = -\left(\frac{\partial V}{\partial T}\right)_P , \tag{3.29}$$

we can find out how S changes when the pressure changes.

Finally, we remark on some relations for an ideal gas that can be derived from such equations. For an ideal gas, $PV = nRT$. From that equation, the right-hand side of (3.28) is equal to $nR/V = P/T$, and we obtain $(\partial S/\partial V)_T = P/T$. On the other hand, the pressure is defined by $(\partial S/\partial V)_E = P/T$. This gives $(\partial S/\partial V)_T = (\partial S/\partial V)_E$ for an ideal gas. This is a consequence of the special characteristics of an ideal gas that the temperature is proportional to the energy and that the energy is uniquely determined by the temperature.

Exercise 7. Calculate by how much the density of oxygen is decreased at the summit of Mount Everest (altitude 8848 m).

Exercise 8. Show that the constant C in (3.14) is the heat capacity of the system.

Exercise 9. Derive the Maxwell relation (3.29).

Part II

Elementary Applications

4

Ideal Gases

Here, we shall apply statistical physics to an ideal gas. We calculate the temperature and pressure as defined by statistical physics, and show that these statistical-mechanical definitions give the same temperature and pressure as do the thermodynamic definitions. We also discuss the thermal properties of an ideal gas consisting of diatomic molecules.

4.1 Quantum Mechanics of a Gas Molecule

Suppose that we enclose N molecules in a box of volume V with adiabatic walls, and let them move with arbitrary velocities. If the influence of gravity can be neglected, the molecules continue to move and they become distributed uniformly, as we have seen in Chap. 1.[1] If there is no interaction between the molecules, the speed of every molecule is conserved. In this case it is impossible for the system to iterate through all possible microscopic states. Therefore we allow a weak interaction between the molecules so that every microscopic state can be realized. A model with this property describes an ideal gas, and also describes a real gas in the dilute limit.

We shall investigate this model by the use of statistical physics in this chapter. For this purpose, we must distinguish between microscopic states. This task cannot be accomplished if we are discussing the motion of a molecule using classical mechanics. In classical mechanics, the state of a molecule is determined by its position $\boldsymbol{r}$ and its momentum $\boldsymbol{p}$. These variables vary continuously, and so we cannot count the number of states that have energies between E and $E + \delta E$. The way out of this difficulty is to use quantum mechanics, which is the correct set of laws governing the microscopic world.

The essence of quantum mechanics is that a molecule behaves both as a particle and as a wave. A wave function $\psi(\boldsymbol{r})$, which is a continuous

[1] Note that since there is no energy transfer between the adiabatic walls and the molecules, the energy of a molecule must be conserved in a collision with a wall. The reflection must be perfectly elastic.

function and takes complex values, is associated with a molecule, and the probability of finding the molecule in an infinitesimal volume δV around $\boldsymbol{r}$ is given by $|\psi(\boldsymbol{r})|^2\,\delta V$. The momentum of the molecule $\boldsymbol{p}$ is related to the wave vector $\boldsymbol{k}$ of the wave function by $\boldsymbol{p} = \hbar\boldsymbol{k}$, where $\hbar = h/2\pi \simeq 1.05 \times 10^{-34}$ J s is the Planck constant divided by 2π.[2] The energy of the molecule $E = p^2/2m$ is related to the angular frequency of the wave ω by $E = \hbar\omega$.

To illustrate quantum mechanics, let us consider a one-dimensional system for simplicity. When a molecule moves along the x-axis with a definite momentum $p = \hbar k$, it is described by the following wave function:

$$\begin{aligned}\psi(x) &= A\mathrm{e}^{\mathrm{i}(kx-\omega t)} \\ &= |A|\cos(kx-\omega t+\alpha) + \mathrm{i}|A|\sin(kx-\omega t+\alpha)\,. \end{aligned} \tag{4.1}$$

Here, Euler's formula,

$$\exp(\mathrm{i}x) = \cos x + \mathrm{i}\sin x\,, \tag{4.2}$$

has been used, and the coefficient A is assumed to be a complex number of the form $A = |A|\mathrm{e}^{\mathrm{i}\alpha}$. This state, with a definite momentum, is a spatially extended state, and the probability of finding a molecule is independent of the position: $|\psi(x)|^2 = |A|^2$. This state is not appropriate for discussing molecules in a box.

To discuss molecules in a box, we need to use standing-wave states. The description of a state in classical and quantum mechanics is illustrated in Fig. 4.1. In classical mechanics, molecules are reflected from the walls, and their momentum changes sign when this happens, from p to $-p$. Likewise, a molecule in a box is described by a wave that is a superposition of waves with momenta p and $-p$. If the size of the box in the x-direction is L_x and the molecule can move in the region $0 < x < L_x$, the wave function is

$$\psi(x,t) = A\sin\left(\frac{px}{\hbar}\right)\mathrm{e}^{-\mathrm{i}\omega t} = \frac{A}{2\mathrm{i}}\left(\mathrm{e}^{\mathrm{i}(px/\hbar-\omega t)} - \mathrm{e}^{-\mathrm{i}(px/\hbar+\omega t)}\right)\,. \tag{4.3}$$

In order to satisfy the boundary condition that $\psi(0,t) = \psi(L_x,t) = 0$, p must be equal to $p_n = nh/2L_x$, where n is a positive integer.[3] In this way, the momentum in the x-direction is restricted to discrete values p_n when the molecule is in a box. This restriction is called *quantization of momentum.* Owing to this quantization, the kinetic energy in the x-direction is also quantized:

$$E_{x,n} = \frac{p_x^2}{2m} = \frac{h^2}{2m}\left(\frac{n}{2L_x}\right)^2\,. \tag{4.4}$$

[2] The Planck constant h is equal to $6.6260693 \times 10^{-34}$ J s. This constant divided by 2π, $\hbar = h/2\pi = 1.05457168 \times 10^{-34}$ J s, is also used frequently.

[3] There is no probability of finding the molecule outside the box. Thus, $\psi(x,t) = 0$ for $x < 0$ and $x > L_x$. The continuity of the wave function requires the boundary condition specified here.

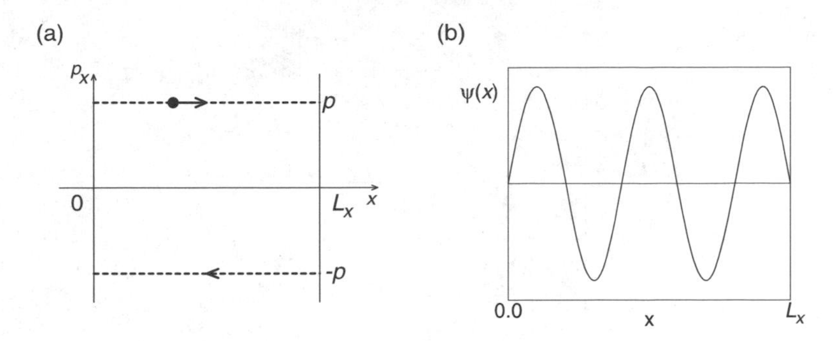

Fig. 4.1. Motion of a molecule in a box. (**a**) Classical mechanics: a molecule of momentum p is reflected at a wall so that after reflection it has a momentum $-p$. (**b**) Wave function in quantum mechanics: the wave function $\psi(x)$ is a continuous function and has a nonzero value only inside the box. In this figure, only the x-component of the motion is depicted, for simplicity

Each of these states, indexed by an integer n, is a microscopic state in which a molecule can be accommodated.

The extension of this standing-wave state to a molecule in a three-dimensional box of size $L_x \times L_y \times L_z$ is straightforward. The wave function and the allowed momenta are

$$\psi(\boldsymbol{r}, t) = A \sin\left(\frac{p_x x}{\hbar}\right) \sin\left(\frac{p_y y}{\hbar}\right) \sin\left(\frac{p_z z}{\hbar}\right) \mathrm{e}^{-\mathrm{i}\omega t} \tag{4.5}$$

and

$$\boldsymbol{p} = (p_x, p_y, p_z) = \left(\frac{nh}{2L_x}, \frac{mh}{2L_y}, \frac{lh}{2L_z}\right), \tag{4.6}$$

where n, m, and l are positive integers.

4.2 Phase Space and the Number of Microscopic States

In order to count the number of microscopic states of an N-molecule system, we need to depict them. For this purpose, we introduce what is called *phase space*. The phase space for a single molecule is the space spanned by the real-space coordinate $\boldsymbol{r}$ and the momentum $\boldsymbol{p}$. If we consider only one-dimensional motion along the x-axis, the phase space becomes two-dimensional, and a point in that space has coordinates (x, p_x). What is shown in Fig. 4.1a is this two-dimensional phase space. In classical mechanics, the state of a molecule at a given time is indicated by a point in this space, and the molecule moves in this space along an orbit, shown by the dashed line, as time elapses. On the other hand, in quantum mechanics, a standing-wave state is represented by a pair of parallel lines at $p_x = \pm nh/2L_x$ in this space, as shown in Fig. 4.2. The nth state has a momentum $p_{x,n} = hn/2L_x$, and the pair of lines enclose an area nh

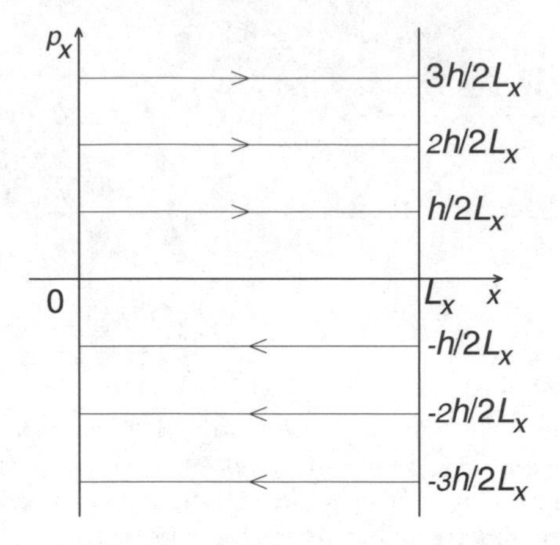

Fig. 4.2. Quantization of the momentum. The state of a molecule in quantum mechanics is described by *a pair of lines* at $p_x = \pm nh/2L_x$ in phase space. The *arrows* indicate that each state is a superposition of right-going and left-going plane waves

in the phase space. Since there are n states in this area, we can conclude that an area of h in this phase space contains just one microscopic state of a molecule.[4]

We have derived this relation on the basis of an inspection of the standing-wave states. However, it is actually a general consequence of Heisenberg's uncertainty principle, which tells us that there is uncertainty in both the position and the momentum of a particle; the position and momentum can only be determined subject to the condition that $\Delta x\, \Delta p \geq h$. In the case of a standing-wave state, the molecule can be anywhere in the box, so that $\Delta x = L_x$. Therefore, $\Delta p \geq h/L_x$ must be satisfied. This is the reason that the lowest momentum is $\pm h/2L_x$, which is nonzero and has uncertainty of $\Delta p = h/L_x$. A spatially more localized wave function can be constructed in the form of a wave packet by superposition of wave functions. If we mix waves with momenta from p to $p + \delta p$, we obtain a wave which has a large amplitude between x and $x + \delta x$ and with a minimum δx determined by δp from $\delta x\, \delta p > h$, as shown in Fig. 4.3. The Heisenberg uncertainty relation is a consequence of this general property of a wave. A wave created in this way occupies an area of $\Delta x\, \Delta p$ in phase space. Thus, a microscopic state of one-dimensional motion can be considered to occupy an area h in two-dimensional phase space, as shown in Fig. 4.4.

This result can be generalized to motions in real three-dimensional space. In this case the phase space becomes six-dimensional, and a point in this

[4] There are, for example, three states in the area between the lines for $n = 3$, with $p_x = \pm\, h/2L_x$, $\pm\, 2h/2L_x$, and $\pm\, 3h/2L_x$, as shown in Fig. 4.2.

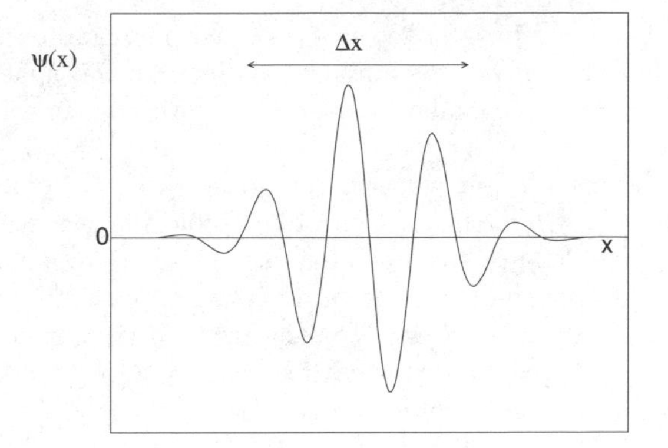

Fig. 4.3. Wave packet. By superposition of waves with momenta in the range $p \pm h/(2\,\Delta x)$, we can create a wave packet of length Δx

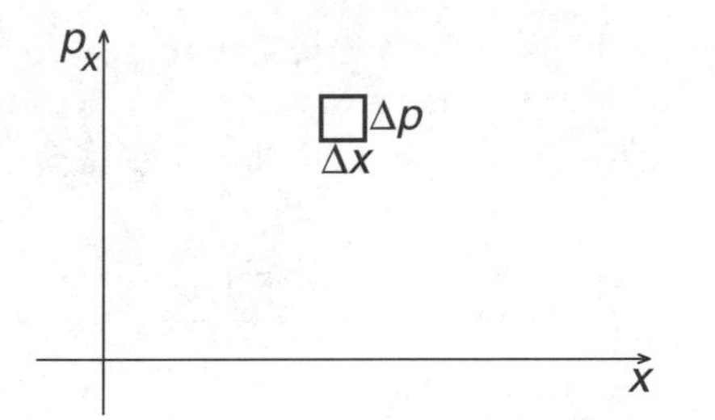

Fig. 4.4. There is one microscopic state in each area of size $\Delta x\,\Delta p = h$ in phase space

space is described by a vector (x, y, z, p_x, p_y, p_z). A quantum mechanical state of a molecule occupies a volume $\Delta x\,\Delta y\,\Delta z\,\Delta p_x\,\Delta p_y\,\Delta p_z = h^3$ in this space.

In statistical physics, we consider a system of N molecules. A microscopic state in this case is depicted in a phase space of $6N$ dimensions. A point in this space has coordinates $(x_1, y_1, z_1, p_{1x}, p_{1y,}, p_{1z}, x_2, y_2, z_2, \ldots)$; that is, six dimensions are assigned to the real-space coordinates and momenta of each molecule. In classical mechanics, a point in this space describes perfectly the state of all the molecules. In quantum mechanics, a microscopic state occupies a volume of h^{3N} in this $6N$-dimensional space.

4.3 Entropy of an Ideal Gas

Once we can count $W(E)$, which is the number of microscopic states that have energies between E and $E+\delta E$, we can obtain the entropy from $S(E) = k_B \ln W(E)$. We can also use the number of states $\Omega_0(E)$ that have energies less than E instead of $W(E)$. In that case,

$$S(E) = k_B \left[\ln \Omega_0 (E) + O(1)\right] . \tag{4.7}$$

For N molecules, the phase space is $6N$-dimensional, and the states with a total energy less than E occupy a nonzero volume in this space. The number of states $\Omega_0(E)$ can be calculated by dividing the volume of phase space by h^{3N}.

We begin the calculation with the simplest case. We consider a system of a single molecule where the motion of the molecule is restricted to a one-dimensional space of le ngth L_x. The phase space is two-dimensional, as shown in Fig. 4.5a. If the energy of the system is E, the momentum of the molecule must satisfy $|p_x| = \sqrt{2mE}$, where m is the mass of the molecule. Therefore, the region of phase space for states with an energy less than E is the shaded area in the figure, and the number of states is

$$\Omega_0 = \frac{2 \times \sqrt{2mE} L_x}{h} . \tag{4.8}$$

Next, the momentum subspace for a molecule in a two-dimensional space of area $L_x \times L_y$ is shown in Fig. 4.5b, and the region corresponding to energies less than E is the shaded circle, the area of which is $\pi(\sqrt{2mE})^2$. Thus,

$$\Omega_0 (E) = \frac{\pi \left(\sqrt{2mE}\right)^2 L_x L_y}{h^2} . \tag{4.9}$$

Similarly, in a three-dimensional space of volume V, the region of the momentum subspace where the energy is less than E is a sphere of radius $\sqrt{2mE}$, and Ω_0 is given by

$$\Omega_0 (E) = \frac{4\pi}{3} \frac{(2mE)^{3/2}}{h^3} V . \tag{4.10}$$

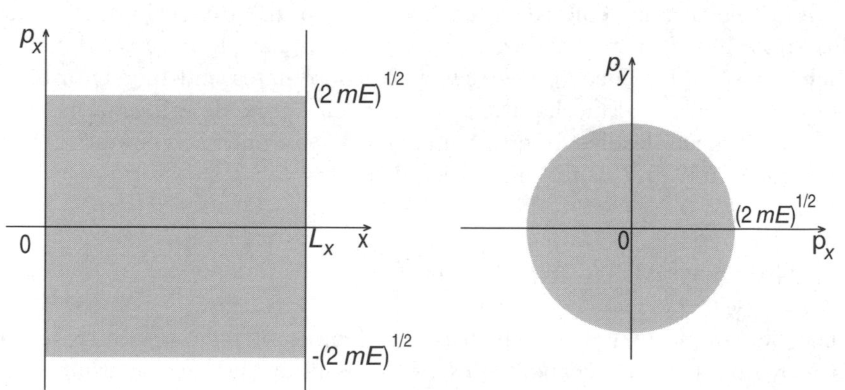

Fig. 4.5. The *shaded area* shows the region in phase space in which the energy of the system is less than E. (**a**) One-dimensional system. (**b**) Two-dimensional system, where only the momentum subspace is shown

Finally, we consider N molecules in a three-dimensional space of volume V. In this case the coordinate part of the phase space has a volume V^N. The momentum part of the phase space is $3N$-dimensional, and the region corresponding to energies less than E is the inside of a hypersphere of radius $\sqrt{2mE}$, namely the region defined by

$$E \geq \sum_{i=1}^{N} \frac{1}{2m} \left(p_{ix}^2 + p_{iy}^2 + p_{iz}^2 \right) . \tag{4.11}$$

The volume of this hypersphere is calculated in Appendix D, and is given by

$$\frac{\pi^{3N/2}}{\Gamma\left(3N/2+1\right)} \left(\sqrt{2mE}\right)^{3N} , \tag{4.12}$$

where $\Gamma(n+1) = n!$ is the gamma function. The number of states $\Omega_0(E)$ is then given by

$$\Omega_0\left(E\right) = \frac{V^N}{h^{3N}} \frac{\pi^{3N/2}}{\Gamma\left(3N/2+1\right)} \left(\sqrt{2mE}\right)^{3N} \frac{1}{N!} . \tag{4.13}$$

In this equation, the final factor $1/N!$ arises from the fact that the molecules cannot be distinguished. That is, the microscopic state in which the ith and jth particles are in the kth and lth quantum mechanical states, respectively, is the same as that in which the ith and the jth particle have been exchanged and are in the lth and kth quantum mechanical states, respectively, Therefore, a microscopic state in which N molecules are in N quantum states α_1, α_2, $\cdots$, α_N, where the ith molecule is in the state α_i, and a microscopic state in which the ith molecule is in the state $\alpha_{P(i)}$, where $P(i)$ is any of the $N!$ permutations of the molecules, are the same microscopic state of the ideal gas. To avoid counting the same microscopic states many times, we need the factor $N!$. We shall come back to this point in Chap. 10. Without this factor $N!$, the entropy has a strange N dependence, which is known as the Gibbs paradox (see also Exercise 11 at the end of this chapter).

Now the entropy of an ideal gas is obtained:

$$\begin{aligned} S\left(E,V,N\right) &= k_{\mathrm{B}} \ln \Omega_0\left(E\right) \\ &= k_{\mathrm{B}} \ln \left(\frac{\left(2\pi m\right)^{3N/2}}{N! h^{3N} \Gamma\left(3N/2+1\right)} \right) + k_{\mathrm{B}} N \ln V + \frac{3k_{\mathrm{B}}N}{2} \ln E \\ &= N k_{\mathrm{B}} \left\{ \ln \frac{V}{N} + \frac{3}{2} \ln \frac{2E}{3N} + \ln \frac{\left(2\pi m\right)^{3/2} \mathrm{e}^{5/2}}{h^3} \right\} . \end{aligned} \tag{4.14}$$

In the final form, Stirling's formula ($\ln N! = N \ln N - N$) has been used. The entropy is proportional to the number of molecules for a fixed number density N/V and fixed energy density $E/V = (E/N)(N/V)$, as it should be.

Before closing this section, we examine the validity of using $\Omega_0(E)$ instead of $W(E)$. The latter is the number of microscopic states with energies between E and $E + \delta E$. Therefore, it is proportional to the surface area of a $3N$-dimensional hypersphere of radius $\sqrt{2mE}$. This surface area is given by the derivative of the volume with respect to the radius, as shown in Appendix D. We obtain

$$W(E) \simeq \frac{V^N}{h^{3N}} \frac{3mN\pi^{3N/2}}{\Gamma(3N/2+1)} (2mE)^{3N/2-1}\, \delta E \frac{1}{N!} \,. \tag{4.15}$$

The difference between the entropy obtained from this $W(E)$ and from (4.14) is negligibly small: apart from an unimportant constant term, the results differ only in that the coefficients of $\ln E$ are $(3N/2 - 1)k_\mathrm{B}$ and $(3N/2)k_\mathrm{B}$, respectively. As N is a macroscopic number, these values for the entropy can be regarded as exactly the same.

4.4 Pressure of an Ideal Gas: Quantum Mechanical Treatment

In Chap. 1, we derived the pressure of an ideal gas as $P = 2E/3V$ using classical mechanics. However, molecules are governed by quantum mechanics, not by classical mechanics. The use of quantum mechanics to describe the state of a molecule is essential for statistical physics. Therefore, we need to calculate the pressure quantum mechanically, and verify that we obtain the same result. The pressure obtained quantum mechanically is compared with the pressure obtained from the entropy in the next section. There, we show agreement between these two pressures.

In the situation shown in Fig. 4.6, the gas is pushing against a piston, and the resulting force is balanced by an external force F. When the balance is infinitesimally broken so that the gas expands, work is extracted from the gas. The extracted work makes the internal energy of the gas decrease. From the decrease in the internal energy as the gas expands slowly, we can calculate the force F and the pressure.

We take the x-axis in the direction in which the piston moves. We consider a situation in which the initial length of the cylinder is L_x, and the cross section of the cylinder is a rectangle of area $L_y \times L_z$. Each molecule is in one of the standing-wave states (4.5), with a momentum given by (4.6). The energy of a molecule in this case is

$$E_{n,m,l} = \frac{h^2}{2m} \left\{ \left(\frac{n}{2L_x}\right)^2 + \left(\frac{m}{2L_y}\right)^2 + \left(\frac{l}{2L_z}\right)^2 \right\} , \tag{4.16}$$

where n, m, and l are positive integers. When we change the length in the x-direction slowly from L_x to $L_x + \delta L_x$, the wave function changes so that

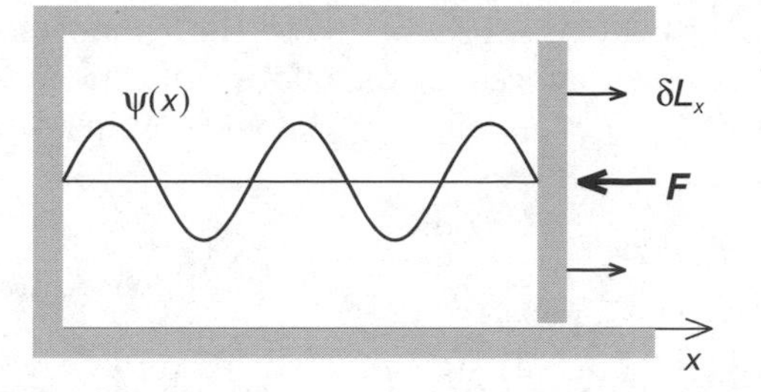

Fig. 4.6. A gas contained in a cylinder with a piston. The piston is pushed by a force F from outside to keep the gas at a given volume. The internal energy of the gas decreases when it expands. When the piston is moved in the x-direction, the wave function $\psi(x)$ is elongated in the x-direction and the energy of the state decreases

the boundary condition that $\psi(\boldsymbol{r}) = 0$ at the wall is maintained. This is accomplished by a slow change in p_x. When L_x increases, $p_x = nh/2L_x$ decreases, and the energy of each molecule decreases. The energy decreases by

$$-\delta E_{n,m,l} = -\frac{\partial E_{n,m,l}}{\partial L_x}\delta L_x = \frac{2}{L_x}\frac{h^2}{2m}\left(\frac{n}{2L_x}\right)^2 \delta L_x\,. \tag{4.17}$$

We equate this decrease to a quantity of work:

$$w = f_x\,\delta L_x\,. \tag{4.18}$$

The force f_x that the molecule exerts on the piston is then obtained:

$$f_x = \frac{2}{L_x}\frac{h^2}{2m}\left(\frac{n}{2L_x}\right)^2 = \frac{2}{L_x} \times (\text{kinetic energy in the } x\text{-direction})\,. \tag{4.19}$$

Summing over all the molecules, and using the fact that the total kinetic energy in the x-direction is one-third of the total kinetic energy E owing to the isotropy of the system, we obtain the force exerted on the piston by the gas:

$$F_x = \frac{2}{3L_x} \times E\,. \tag{4.20}$$

From this force, we obtain the pressure as $P = F_x/(L_yL_z) = (2/3)E/V$. This is the same result as that obtained from classical mechanics. The total kinetic energy E is nothing but the internal energy of the (ideal) gas.

4.5 Statistical-Mechanical Temperature and Pressure

In this section, we show that the temperature and pressure defined by statistical mechanics are the same as those defined by thermodynamics. In order to avoid confusion, we add a subscript "s" to the symbols for these quantities defined by statistical mechanics, and therefore write them as T_s and P_s. From their definitions and the expression for the entropy in (4.14), T_s and P_s are given by

$$\frac{1}{T_\mathrm{s}} = \left(\frac{\partial S}{\partial E}\right)_{V,N} = \frac{3}{2} N k_\mathrm{B} \frac{1}{E}, \tag{4.21}$$

$$\frac{P_\mathrm{s}}{T_\mathrm{s}} = \left(\frac{\partial S}{\partial V}\right)_{E,N} = N k_\mathrm{B} \frac{1}{V}. \tag{4.22}$$

By eliminating T_s from these equations, we obtain $P_\mathrm{s} = (2/3)E/V$. Therefore, P_s is the same as the pressure defined mechanically, which was calculated in the previous section.

Next we examine the temperature. The thermodynamic absolute temperature is defined through the efficiency of an ideal heat engine. This definition is independent of any material. Now, thermodynamics tells us that in the case of an ideal gas whose internal energy depends only on the temperature and whose state of equation is given by the Boyle–Charles law $PV = nRT = Nk_\mathrm{B}T$, the T in the right-hand side of that equation coincides with the absolute temperature. The ideal gas treated in this chapter satisfies this necessary condition. Namely, (4.21) shows that E depends only on T_s, and (4.22) tells us that $PV = k_\mathrm{B}NT_\mathrm{s}$. Therefore, the temperature defined by statistical physics coincides with the thermodynamic absolute temperature.

From this comparison, we have found that the statistical-mechanical temperature and pressure of an ideal gas are the same as the temperature and pressure obtained from thermodynamics, as expected. This leads us to the conclusion that the definitions of temperature and pressure in statistical physics are equivalent to the corresponding definitions in thermodynamics for any system. This can be proved by the following consideration. Let an arbitrary system I be in thermal and mechanical equilibrium with an ideal gas, which we shall call system II. Then, as explained in Sect. 2.3, the principle of equal probability makes T_s and P_s the same for these systems, i.e. $T_\mathrm{s}^{(\mathrm{I})} = T_\mathrm{s}^{(\mathrm{II})}$ and $P_\mathrm{s}^{(\mathrm{I})} = P_\mathrm{s}^{(\mathrm{II})}$. On the other hand, the thermodynamic temperature and the mechanical pressure are also the same for the two systems, i.e. $T^{(\mathrm{I})} = T^{(\mathrm{II})}$ and $P^{(\mathrm{I})} = P^{(\mathrm{II})}$. Since $T_\mathrm{s}^{(\mathrm{II})} = T^{(\mathrm{II})}$ and $P_\mathrm{s}^{(\mathrm{II})} = P^{(\mathrm{II})}$, we conclude that $T^{(\mathrm{I})} = T_\mathrm{s}^{(\mathrm{I})}$ and $P^{(\mathrm{I})} = P_\mathrm{s}^{(\mathrm{I})}$. Namely, the statistical-mechanical temperature and pressure coincide with the thermodynamic temperature and the mechanical pressure for any system.

4.6 Partition Function of an Ideal Gas

Above, we calculated the entropy of an ideal gas using the microcanonical distribution. We can also obtain the same result using the canonical distribution, in which the temperature is fixed by a heat bath. The partition function in this case is

$$\begin{aligned} Z(T,V,N) &= \sum_i \exp(-\beta E_i) \\ &= \frac{1}{N!}\int \mathrm{e}^{-\beta E(\boldsymbol{r}_1,\boldsymbol{r}_2,\cdots,\boldsymbol{p}_1,\boldsymbol{p}_2,\cdots)} \frac{\mathrm{d}^3\boldsymbol{r}_1\,\mathrm{d}^3\boldsymbol{p}_1\cdots}{h^{3N}} . \end{aligned} \tag{4.23}$$

Here, $\beta = 1/k_\mathrm{B}T$.[5] The $6N$-dimensional phase space of N molecules has been reduced by a factor of $1/N!$ by considering the exchange of molecules, and has been divided into cells of volume h^{3N} corresponding to microscopic states. The energy of a microscopic state, $E(\boldsymbol{r}_1,\boldsymbol{r}_2,\cdots,\boldsymbol{p}_1,\boldsymbol{p}_2,\cdots)$ does not depend on the position $\boldsymbol{r}_i$:

$$E = \frac{1}{2m}\left(\boldsymbol{p}_1{}^2 + \boldsymbol{p}_2{}^2 + \boldsymbol{p}_3{}^2\cdots\right) . \tag{4.24}$$

The integral over the space coordinates gives V^N, and the integral over the momentum gives

$$\begin{aligned} Z(T,V,N) &= \frac{V^N}{N!}\frac{1}{h^{3N}}\int \mathrm{d}^3\boldsymbol{p}_1 \exp\left(-\beta\frac{p_1^2}{2m}\right)\int \mathrm{d}^3\boldsymbol{p}_2 \exp\left(-\beta\frac{p_2^2}{2m}\right)\cdots \\ &= \frac{1}{N!}\frac{V^N}{h^{3N}}\left(\frac{2\pi m}{\beta}\right)^{3N/2} \\ &\simeq \left(\frac{2\pi m k_\mathrm{B}T}{h^2}\right)^{3N/2}\left(\frac{V}{N}\mathrm{e}\right)^N . \end{aligned} \tag{4.25}$$

From this partition function, the free energy F, the internal energy U, and the entropy S can be calculated:

$$\begin{aligned} F(T,V,N) &= -k_\mathrm{B}T\ln Z \\ &= -\frac{3Nk_\mathrm{B}T}{2}\ln\left(\frac{2\pi m k_\mathrm{B}T}{h^2}\right) - Nk_\mathrm{B}T\ln\left(\frac{V\mathrm{e}}{N}\right) , \end{aligned} \tag{4.26}$$

$$\begin{aligned} U(T,V,N) &= -\left(\frac{\partial}{\partial\beta}\ln Z\right)_{V,N} = -\frac{\partial}{\partial\beta}\left[\frac{3N}{2}\ln\left(\frac{1}{\beta}\right)\right] \\ &= \frac{3}{2}Nk_\mathrm{B}T , \end{aligned} \tag{4.27}$$

$$\begin{aligned} S(T,V,N) &= -\left(\frac{\partial F}{\partial T}\right)_{V,N} \\ &= \frac{3}{2}Nk_\mathrm{B}\ln\left(\frac{2\pi m k_\mathrm{B}T}{h^2}\right) + Nk_\mathrm{B}\ln\left(\frac{V\mathrm{e}}{N}\right) + \frac{3}{2}Nk_\mathrm{B} . \end{aligned} \tag{4.28}$$

As $T \propto U = E$, this entropy coincides with (4.14).

[5] In statistical physics, β is always used for $1/k_\mathrm{B}T$.

We can also verify that the pressure obtained from the free energy is the same as before:

$$P = -\left(\frac{\partial F}{\partial V}\right)_{T,N} = Nk_{\rm B}T\frac{1}{V}\,. \tag{4.29}$$

From the internal energy, we can calculate the heat capacity at constant volume:

$$C_{\rm V} = \left(\frac{\partial U}{\partial T}\right)_{V,N} = \frac{3}{2}Nk_{\rm B}\,. \tag{4.30}$$

Since the internal energy U depends only on the temperature, and $(\partial V/\partial T)_P = V/T$, (1.5) tells us that the constant-pressure heat capacity $C_{\rm P}$ is given by

$$C_{\rm P} = C_{\rm V} + Nk_{\rm B}\,. \tag{4.31}$$

This relation is known as the Mayer relation. (see Exercise 4 in Chap. 1.)

The ratio $C_{\rm P}/C_{\rm V} \equiv \gamma$ is known as the specific-heat ratio. In the present case $\gamma = 1.67$. This value coincides with the experimental value for helium gas, but for air at room temperature, γ is equal to 1.403. This discrepancy arises from the fact that air consists of diatomic molecules, as explained in the next section.

4.7 Diatomic Molecules

4.7.1 Decomposition of the Partition Function

The monatomic molecules that we have considered in the previous sections have three degrees of freedom, associated with the motion of the center of gravity. On the other hand, a diatomic molecule has six degrees of freedom, since it consists of two atoms. Therefore, in addition to the three degrees of freedom associated with the center of gravity, it has three degrees of freedom associated with its internal motion. To describe the internal motion, we take a coordinate system attached to the molecule as shown in Fig. 4.7. The three degrees of freedom of the internal motion can then be divided into two kinds:

- *Vibration* in the x-direction. This has one degree of freedom.
- *Rotation* around the y- and z-axes. This has two degrees of freedom.

To be exact, there are also degrees of freedom associated with the motion of the electrons and with the rotation and vibration of the nucleus, but these degrees of freedom need not be considered at room temperature, for a reason that will be explained later. So here we consider the statistical mechanics of diatomic molecules taking account of the above six degrees of freedom.

Now, because of these internal motions, the calculation of the partition function of a diatomic-molecule gas becomes a little complicated. To reduce the complexity, we shall first show that the total partition function can be

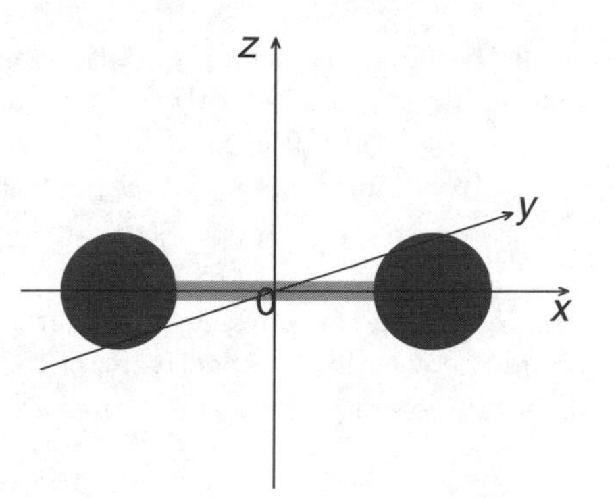

Fig. 4.7. Diatomic molecule, and coordinate system attached to the molecule

obtained from a product of single-molecule partition functions in the case of an ideal gas. This is because there is no interaction, and each molecule is independent of the others. Let us write the energy of the jth quantum mechanical state of the ith molecule as $E_i(j)$, where $0 \le j \le \infty$ and $0 \le i \le N$. When the ith molecule is in the j_ith state, the total energy of the gas $\mathcal{E}$ is

$$\mathcal{E}(j_1, j_2, \cdots, j_N) = \sum_i E_i(j_i) . \tag{4.32}$$

The partition function is then

$$\begin{aligned} Z &= \frac{1}{N!} \sum_{j_1} \sum_{j_2} \cdots \sum_{j_N} \exp\left[-\beta \mathcal{E}(j_1, j_2, \cdots, j_N)\right] \\ &= \frac{1}{N!} \sum_{j_1} \sum_{j_2} \cdots \sum_{j_N} \prod_i \exp\left[-\beta E_i(j_i)\right] \\ &= \frac{1}{N!} \prod_i \left\{ \sum_{j_i} \exp\left[-\beta E_i(j_i)\right] \right\} \\ &= \frac{1}{N!} \left\{ \sum_j \exp\left[-\beta E_1(j)\right] \right\}^N . \end{aligned} \tag{4.33}$$

In the last line, the function in the large braces is the partition function of the first molecule z_1,

$$z_1 \equiv \sum_j \exp\left(-\beta E_1(j)\right) . \tag{4.34}$$

Since all the molecules are the same, $z_1 = z_2 = \cdots = z_N$. Thus the partition function of an ideal gas can be written as a product of single-molecule partition functions. It is easy to check that the partition function of a monatomic ideal gas (4.25) can be obtained from this equation.

For the calculation of the single-molecule partition function, it is a good approximation to consider the six degrees of freedom independently. In this case the energy of a microscopic state of a molecule is given by the sum of the energies for each degree of freedom. Thus the energy of the ith molecule is

$$E_i = E_i^{(\mathrm{CG})} + E_i^{(\mathrm{V})} + E_i^{(\mathrm{R})}, \tag{4.35}$$

where $E_i^{(\mathrm{CG})}$, $E_i^{(\mathrm{V})}$, and $E_i^{(\mathrm{R})}$ are the energies associated with the center of gravity, the vibration, and the rotation, respectively, of the ith molecule. Since the energy is the sum of the energies for each degree of freedom, z_1 can be rewritten as follows:

$$\begin{aligned} z_1 &= \sum_j \exp\left[-\beta E_1(j)\right] \\ &= \sum_j \exp\left[-\beta E_1^{(\mathrm{CG})}(j)\right] \sum_k \exp\left[-\beta E_1^{(\mathrm{V})}(k)\right] \sum_j \exp\left[-\beta E_1^{(\mathrm{R})}(l)\right] \\ &\equiv Z^{(\mathrm{CG})} Z^{(\mathrm{V})} Z^{(\mathrm{R})}. \end{aligned} \tag{4.36}$$

The total partition function and the free energy are then

$$Z = \frac{1}{N!}\left\{Z^{(\mathrm{CG})} Z^{(\mathrm{V})} Z^{(\mathrm{R})}\right\}^N \tag{4.37}$$

and

$$\begin{aligned} F &= -k_\mathrm{B} T N \left\{\ln Z^{(\mathrm{CG})} + \ln Z^{(\mathrm{V})} + \ln Z^{(\mathrm{R})}\right\} + k_\mathrm{B} T \ln N! \\ &= F^{(\mathrm{CG})} + F^{(\mathrm{V})} + F^{(\mathrm{R})} + k_\mathrm{B} T N \ln N. \end{aligned} \tag{4.38}$$

In the following, we calculate each term in this equation.

4.7.2 Center-of-Gravity Part: $Z^{(\mathrm{CG})}$

This part is the same as that for a monatomic molecule. Writing the center coordinate, the total momentum, and the total mass as $\boldsymbol{R}$, $\boldsymbol{P}$, and $M = 2m$, respectively, we can calculate this part as follows:

$$\begin{aligned} Z^{(\mathrm{CG})} &= \sum_j \exp\left[-\beta E^{(\mathrm{CG})}(j)\right] = \frac{1}{h^3}\int \mathrm{d}^3\boldsymbol{R}\,\mathrm{d}^3\boldsymbol{P} \exp\left(-\beta\frac{P^2}{2M}\right) \\ &= \frac{V}{h^3}\left(2\pi M k_\mathrm{B} T\right)^{3/2}. \end{aligned} \tag{4.39}$$

The free energy and the internal energy for this part of the partition function are

$$F^{(\mathrm{CG})} = -k_\mathrm{B} T N \ln Z^{(\mathrm{CG})} = -\frac{3}{2} k_\mathrm{B} T N \ln\left(2\pi M k_\mathrm{B} T\right) \tag{4.40}$$

and

$$U^{(\mathrm{CG})} = -N\frac{\partial}{\partial\beta}\ln Z^{(\mathrm{CG})} = \frac{3}{2} N k_\mathrm{B} T. \tag{4.41}$$

4.7.3 Vibrational Part: $Z^{(\mathrm{V})}$

To discuss vibration, we consider a diatomic molecule as two atoms of mass m connected by a spring of spring constant k, as shown in Fig. 4.8. The equation of motion for this system in classical mechanics is

$$\mu \ddot{x} = -k(x-d)\,, \tag{4.42}$$

where $\mu = m/2$ is the reduced mass, x is the distance between the atoms, and d is the mean value of this distance. This is the equation of motion of a harmonic oscillator, and the solution is $x = A\sin(\omega t + \alpha) + d$, where $\omega = \sqrt{k/\mu}$. The energy is the sum of the kinetic energy

$$\frac{\mu}{2}\dot{x}^2 = \frac{k}{2}A^2\cos^2(\omega t + \alpha) \tag{4.43}$$

and the potential energy

$$\frac{1}{2}k(x-d)^2 = \frac{k}{2}A^2\sin^2(\omega t + \alpha)\,, \tag{4.44}$$

and we obtain $E^{(\mathrm{V})} = (k/2)A^2$. The amplitude A can take any value, and so the energy is continuous. We cannot count the number of microscopic states in classical mechanics.

On the other hand, in quantum mechanics, the energy of a harmonic oscillator is given by

$$E^{(\mathrm{V})} = \left(n + \frac{1}{2}\right)h\frac{\omega}{2\pi} = \left(n + \frac{1}{2}\right)\hbar\omega\,, \tag{4.45}$$

where $n = 0, 1, 2, \cdots$ is a nonnegative integer. The lowest-energy state, the ground state, is obtained when $n = 0$. Even in this case, the energy is nonzero. This is because a zero-energy state is not allowed by the uncertainty principle: $E^{(\mathrm{V})} = 0$ would mean that there was no vibration, and so the distance between the atoms would be fixed at d without any fluctuation, which would mean that the relative momentum must be completely uncertain, i.e. $\Delta p \to \infty$. This means that the energy would be very large, in contradiction to the initial assumption that $E^{(\mathrm{V})} = 0$. Therefore, even in the ground state, there must be some vibration. This vibration is known as the *zero-point vibration.*

We cannot give a derivation of (4.45) here; that belongs in a course on quantum mechanics. Here we shall simply accept the result and calculate the

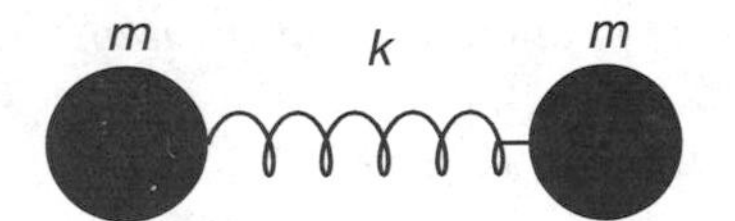

Fig. 4.8. Model of a diatomic molecule. To discuss vibration, we consider a diatomic molecule as two atoms of mass m connected by a spring of spring constant k

partition function. The partition function can be expressed in terms of the hyperbolic function $\sinh x$:[6]

$$Z^{(\mathrm{V})} = \sum_{n=0}^{\infty} \exp\left[-\beta\left(n+\frac{1}{2}\right)\hbar\omega\right] = \frac{\mathrm{e}^{-\beta\hbar\omega/2}}{1-\mathrm{e}^{-\beta\hbar\omega}} = \frac{1}{2\sinh\left(\beta\hbar\omega/2\right)} . \quad (4.46)$$

The free energy and internal energy associated with vibration can then be obtained:

$$F^{(\mathrm{V})} = -k_{\mathrm{B}}TN\ln Z^{(\mathrm{V})} = k_{\mathrm{B}}TN\ln\left[2\sinh\left(\frac{1}{2}\beta\hbar\omega\right)\right] \quad (4.47)$$

and

$$\begin{aligned} U^{(\mathrm{V})} &= -N\frac{\partial}{\partial\beta}\ln Z^{(\mathrm{V})} = N\frac{\partial}{\partial\beta}\ln\left[2\sinh\left(\frac{1}{2}\beta\hbar\omega\right)\right] \\ &= \frac{N}{2}\hbar\omega\coth\left(\frac{1}{2}\beta\hbar\omega\right) \\ &= \frac{N}{2}\hbar\omega + \frac{N\hbar\omega}{\mathrm{e}^{\beta\hbar\omega}-1} . \end{aligned} \quad (4.48)$$

We now examine the temperature dependence of the internal energy. When the temperature is low enough that $\beta\hbar\omega \gg 1$, or $\hbar\omega \gg k_{\mathrm{B}}T$, is satisfied, we have $\mathrm{e}^{\beta\hbar\omega} \gg 1$. In this case the second term can be neglected, and $U^{(\mathrm{V})} = (N/2)\hbar\omega$. This means that every molecule is in the ground state at low temperature. On the other hand, at a temperature high enough that $\beta\hbar\omega \ll 1$, or $\hbar\omega \ll k_{\mathrm{B}}T$, is satisfied, we have $\mathrm{e}^{\beta\hbar\omega} \simeq 1+\beta\hbar\omega$, and $U^{(\mathrm{V})} \simeq (N/2)\hbar\omega + Nk_{\mathrm{B}}T \simeq Nk_{\mathrm{B}}T$. The average energy of a molecule is now $k_{\mathrm{B}}T$; namely, every molecule has an average kinetic energy and an average potential energy of $(1/2)k_{\mathrm{B}}T$ each. The fact that the kinetic energy of a system at high temperature is equal to $(1/2)k_{\mathrm{B}}T$ for each degree of freedom is known empirically and is called the *law of equipartition*, and the energy $(1/2)k_{\mathrm{B}}T$ is often called the *thermal energy*. This law is obeyed by the center-of-gravity part of the internal energy (4.41) also. However, the law of equipartition is obeyed only approximately in the high-temperature limit in the case of vibration. Owing to the temperature dependence of the internal energy, the contribution to the heat capacity from vibration, $C^{(\mathrm{V})} = \left(\partial U^{(\mathrm{V})}/\partial T\right)$, has a temperature dependence of the form shown in Fig. 4.9. Namely, $C^{(\mathrm{V})}$ is exponentially small at low temperature, but approaches Nk_{B} at high temperature.

The temperature dependence described above arises from the discreteness of the energy of vibration $E_n^{(\mathrm{V})} = (n+1/2)\hbar\omega$. The integer n in this equation is an example of a quantity called a *quantum number*. It indicates how much vibrational energy a molecule has in units of $\hbar\omega$. The average vibrational energy of a molecule can be written as

$$\langle E^{(\mathrm{V})}\rangle = (\langle n\rangle + 1/2)\hbar\omega , \quad (4.49)$$

[6] Hyperbolic functions are defined and briefly described in Appendix E.

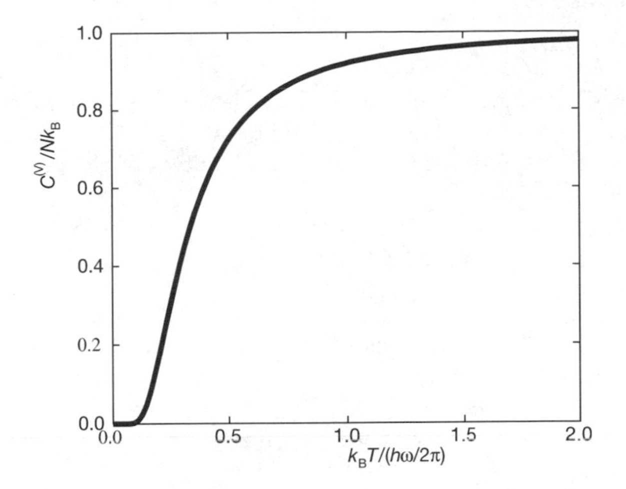

Fig. 4.9. Contribution to the heat capacity from vibration for a diatomic molecule. $C^{\mathrm{V}}/Nk_{\mathrm{B}}$ is plotted as a function of $k_{\mathrm{B}}T/\hbar\omega = k_{\mathrm{B}}T/(h\omega/2\pi)$

where $\langle n \rangle$ is the thermal average of the quantum number n, and is given by

$$\langle n \rangle = \frac{\sum_{n=1}^{\infty} n \exp\left[-\beta E_n^{(\mathrm{V})}\right]}{\sum_{n=0}^{\infty} \exp\left[-\beta E_n^{(\mathrm{V})}\right]} = \frac{1}{\mathrm{e}^{\beta\hbar\omega} - 1} \,. \tag{4.50}$$

The result for $U^{(\mathrm{V})}$ (4.48) can be recovered from these equations, since $U^{(\mathrm{V})} = N\langle E^{(\mathrm{V})}\rangle$. This distribution of n is called the *Bose distribution*. It has already appeared in (2.9), where we considered a distribution of money. There, the unit of energy $\hbar\omega$ was replaced by a unit of money, 1 yen.

When the thermal energy $(1/2)k_{\mathrm{B}}T$ is much smaller than the unit of energy $\hbar\omega$, the system cannot accept energy from a heat bath. In such a case, we can neglect the corresponding degree of freedom, and say that that degree of freedom is "dead"; in this situation, that degree of freedom makes a negligible contribution to any observable. For nitrogen or oxygen, $\hbar\omega/k_{\mathrm{B}} \simeq 2000\,\mathrm{K} \gg T \simeq 300\,\mathrm{K}$, and so the vibrational degree of freedom is dead at 300 K. Likewise, the units of energy for the motion of electrons in a molecule and for the motions within a nucleus are much higher than the thermal energy. This is the reason we can neglect the degrees of freedom associated with those motions. The energy of the center of gravity is also quantized, as can be seen from (4.4). However, the unit of energy in this case, $(1/2m)(h/2L)^2$, is very small for macroscopic values of L, and so the translational motion of the center of gravity is not dead even at low temperature. Thus, the law of equipartition is obeyed in this case down to the lowest temperatures.

4.7.4 Rotational Part: $Z^{(\mathrm{R})}$

First we consider the system using classical mechanics. If the molecule rotates about the z-axis with an angular frequency ω, the velocity is given by $v = (d/2)\omega$, where d is the distance between the centers of the two atoms. Thus the energy is

$$E^{(\mathrm{R})} = 2 \times \frac{m}{2}\left(\frac{d}{2}\omega\right)^2 = \frac{1}{4}md^2\omega^2 = \frac{1}{2}I\omega^2\,, \tag{4.51}$$

where $I = (1/2)md^2$ is the moment of inertia of the molecule. Since ω is a continuous variable, $E^{(\mathrm{R})}$ is also continuous, and we cannot count the number of microscopic states.

The angular momentum associated with this rotation, $\boldsymbol{L} = \boldsymbol{r} \times \boldsymbol{p}$, has only a z-component,

$$L_z = 2 \times \frac{d}{2} \times m\frac{d}{2}\omega = \frac{1}{2}md^2\omega = I\omega\,. \tag{4.52}$$

This is also a continuous variable in classical mechanics, but is quantized in quantum mechanics:

$$L_z = I\omega = l\hbar\,, \qquad l = 0, 1, 2, \cdots. \tag{4.53}$$

In terms of this l, the square of the angular momentum is

$$L^2 = l\,(l+1)\,\hbar^2\,, \tag{4.54}$$

and the energy is

$$E^{(\mathrm{R})} = \frac{1}{2I}L^2 = l\,(l+1)\,\frac{\hbar^2}{2I}\,. \tag{4.55}$$

For each value of l, there are $2l+1$ states owing to the freedom in the choice of the direction of the axis of rotation.[7] If we accept this result of quantum mechanics, the partition function can be calculated as follows:

$$Z^{(\mathrm{R})} = \sum_{l=0}^{\infty}(2l+1)\exp\left\{-l\,(l+1)\,\frac{\beta\hbar^2}{2I}\right\}. \tag{4.56}$$

This summation cannot be done analytically. However, at high temperature it can be done approximately. When $\beta\hbar^2/2I \ll 1$, the exponential factor becomes small only at large l. The factor $(2l+1)$ makes the contributions from large values of l more important than those from small values. In this case it is a good approximation to calculate the summation as an integral. As the variable of the integral, we take

$$\varepsilon\,(l) \equiv l\,(l+1)\,\frac{\beta\hbar^2}{2I}\,. \tag{4.57}$$

[7] When $L^2 = l(l+1)\hbar^2$, L_z takes $2l+1$ quantized values, which can be expressed as $L_z = m_z\hbar$, where m_z is an integer that satisfies $-l \le m_z \le l$.

Then

$$\frac{\mathrm{d}\varepsilon}{\mathrm{d}l} = (2l+1)\,\frac{\beta\hbar^2}{2I}\,, \tag{4.58}$$

and the partition function can be approximated as

$$\begin{aligned} Z^{(\mathrm{R})} &\cong \int_0^\infty \mathrm{d}l\ (2l+1)\exp\left\{-l\,(l+1)\,\frac{\beta\hbar^2}{2I}\right\} \\ &= \frac{2I}{\beta\hbar^2}\int_0^\infty \mathrm{d}\varepsilon\ \exp\left(-\varepsilon\right) = \frac{2I}{\hbar^2}k_{\mathrm{B}}T\,. \end{aligned} \tag{4.59}$$

From this $Z^{(\mathrm{R})}$, the free energy and the internal energy are

$$F^{(\mathrm{R})} \simeq -k_{\mathrm{B}}TN\ln\left(\frac{2I}{\hbar^2}k_{\mathrm{B}}T\right) \tag{4.60}$$

and

$$U^{(\mathrm{R})} \simeq -N\frac{\partial}{\partial\beta}\ln Z^{(\mathrm{R})} = \frac{N}{\beta} = Nk_{\mathrm{B}}T\,. \tag{4.61}$$

This result for the internal energy can again be interpreted as a consequence of the equipartition law. Namely, the thermal energy $(1/2)k_{\mathrm{B}}T$ is given to each of the two degrees of freedom of each molecule. In this limit, the contribution to the heat capacity is

$$C^{(\mathrm{R})} = \frac{\partial U^{(\mathrm{R})}}{\partial T} = k_{\mathrm{B}}N\,. \tag{4.62}$$

For an oxygen molecule, for which $m = 2.66\times 10^{-26}\,\mathrm{kg}$ and $d = 2.8\ \text{Å} = 0.28\times 10^{-9}\,\mathrm{m}$, the moment of inertia I is $1.05\times 10^{-45}\,\mathrm{kg\,m^2}$, and so

$$\frac{1}{k_{\mathrm{B}}}\frac{\hbar^2}{2I} = 0.38\,\mathrm{K}\,. \tag{4.63}$$

Therefore, room temperature can be considered as a high temperature. The situation is similar for a nitrogen molecule. Thus the law of equipartition is obeyed for the rotation of air molecules at room temperature.

The total heat capacity of a diatomic gas is the sum of all of the above contributions:

$$C_{\mathrm{V}} = C^{(\mathrm{CG})} + C^{(\mathrm{V})} + C^{(\mathrm{R})} = \frac{3}{2}k_{\mathrm{B}}N + C^{(\mathrm{V})} + k_{\mathrm{B}}N = \frac{5}{2}k_{\mathrm{B}}N\,. \tag{4.64}$$

This applies at room temperature, where $C^{(\mathrm{V})}$ is negligibly small. Then, from Mayer's relation, we obtain $C_{\mathrm{P}} = C_{\mathrm{V}} + k_{\mathrm{B}}N = (7/2)k_{\mathrm{B}}N$, and the specific-heat ratio γ is equal to $7/5 = 1.4$. This value agrees with experimental results for air.

Exercise 10. Evaluate the lowest kinetic energy of an oxygen molecule in a cubic box of volume $10\,\mathrm{cm}^3$. Express the result in joules and calculate the temperature at which the thermal energy $k_{\mathrm{B}}T$ corresponds to this energy.

Exercise 11. Gibbs noticed that without the factor $N!$, the entropy of an ideal gas has a strange behavior. Obtain an expression for S without this factor. Use this expression and show that the total entropy of two identical systems $S(2E, 2V, 2N)$ is different from $2S(E, V, N)$.

Exercise 12. Calculate the entropy of an ideal gas using (4.15), and compare the result with (4.14).

Exercise 13. The frequency ν of vibration of an O_2 molecule is 4.74×10^{13} Hz. Evaluate the spring constant $k = \mu\omega^2$ of the bond between the oxygen atoms, where $\mu = m/2$ is the reduced mass, $m = 2.66 \times 10^{-26}$ kg, and $\omega = 2\pi\nu$.

5

The Heat Capacity of a Solid, and Black-Body Radiation

The molar heat of solids approach a common value at high temperatures, whereas they decrease in proportion to T^3 at low temperatures. To explain this behavior, we introduce here the Einstein and Debye models as models of the crystal lattices of solids, and investigate them. Black-body radiation has the same mathematical structure as the Debye model. We shall derive Planck's radiation formula.

5.1 Heat Capacity of a Solid I – Einstein Model

In the previous chapter, we showed how we can understand the heat capacity of an ideal gas. Now we shall try to understand the heat capacity of a solid using statistical mechanics. First, we summarize the experimental facts:

- Around room temperature, ordinary simple solids have a molar heat[1] of about $3k_{\mathrm{B}}N \simeq 25\,\mathrm{J\,mol^{-1}\,K^{-1}}$. This fact is called the *Dulong–Petit law*. Some examples are listed in Table 5.1.
- At low temperatures, the molar heat decreases in proportion to T^3.

We shall explain these facts in this and the following section.

In this section, we investigate solids using the Einstein model. Just as the motion of atoms contributes to the heat capacity of a gas, we expect that the motion of atoms or ions will contribute to the heat capacity of a solid. The ideal solid is a crystal, where atoms are arranged periodically in a lattice. In a simple cubic lattice, eight neighboring atoms sit at the corners of a cube. The atoms can move around their equilibrium positions, which are called lattice points. Owing to the surrounding atoms, the potential energy of an atom is lowest at its lattice point. We can depict this situation schematically as one where atoms are connected by springs, as shown in Fig. 5.1. Einstein simplified

[1] The molar heat is the heat capacity per mole.

Table 5.1. Molar heats of solids at 25°C in J mol^{-1} K^{-1}. Note that diamond has an exceptionally small value

Material	Molar heat
Aluminum	24.34
Gold	25.38
Silver	25.49
Iron	25.23
Sulfur	22.60
Phosphorus	23.8
Diamond	6.115

the real situation by assuming that each atom is connected to its lattice point by a spring independently of the other atoms.

In this model, the motion of each atom is a harmonic oscillation. Let the angular frequency of the oscillation be ω in all three directions x, y, and z. Each atom then has an energy

$$E_{n_x,n_y,n_z} = \left(n_x + \frac{1}{2} + n_y + \frac{1}{2} + n_z + \frac{1}{2}\right)\hbar\omega\,, \tag{5.1}$$

where n_x, n_y, and n_z are the quantum numbers of oscillation in the x, y, and z directions, respectively. From this energy, the partition function of a system of N atoms is

$$\begin{aligned} Z &= \left\{\sum_{n_x=0}^{\infty}\sum_{n_y=0}^{\infty}\sum_{n_z=0}^{\infty}\exp(-\beta E_{n_x,n_y,n_z})\right\}^N \\ &= \left\{\sum_{n=0}^{\infty}\exp\left[-\beta\left(n+\frac{1}{2}\right)\hbar\omega\right]\right\}^{3N} \\ &= \left\{\frac{1}{\mathrm{e}^{(1/2)\beta\hbar\omega} - \mathrm{e}^{-(1/2)\beta\hbar\omega}}\right\}^{3N} \\ &= \left\{\frac{1}{2\sinh\left[(1/2)\beta\hbar\omega\right]}\right\}^{3N}, \end{aligned} \tag{5.2}$$

where $\beta = 1/k_{\mathrm{B}}T$ as usual.[2] The internal energy U and the constant-volume heat capacity C_{V} are given by

$$U = -\frac{\partial}{\partial\beta}\ln Z = \frac{3}{2}N\hbar\omega\coth\left(\frac{\hbar\omega}{2k_{\mathrm{B}}T}\right) \tag{5.3}$$

[2] In this equation, we do not have a factor $1/N!$. This is because each atom can be identified by the position of its lattice point.

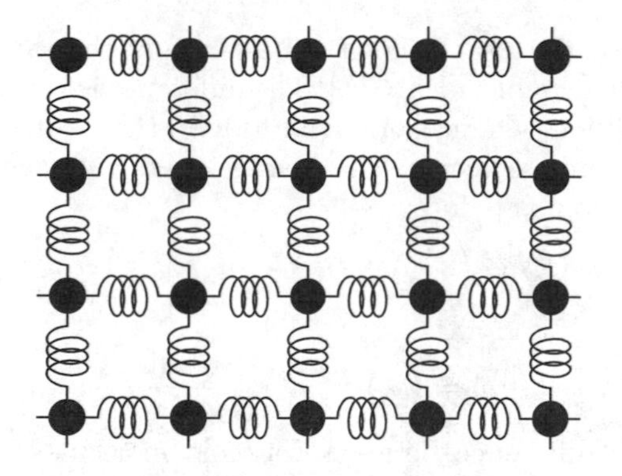

Fig. 5.1. Crystal lattice. Atoms, shown by *solid circles*, are arranged periodically in a lattice. The equilibrium position of each atom is called a lattice point. The atoms can be considered as connected by springs

and

$$C_V = \frac{dU}{dT} = 3Nk_B \left(\frac{\hbar\omega}{2k_BT}\right)^2 \frac{1}{\sinh^2(\hbar\omega/2k_BT)} . \tag{5.4}$$

The temperature dependence of the heat capacity is plotted in Fig. 5.2. When $T \to \infty$, $\hbar\omega/2k_BT \to 0$, and C_V approaches $3Nk_B$. The molar heat in this limit is $3N_Ak_B = 3R = 24.93$ J/mol K. This is the Dulong–Petit law. We can understand this result by using the equipartition law; namely, a thermal energy of $(1/2)k_BT$ is given to the kinetic energy and the potential energy in each of the x, y, and z directions in this limit. On the other hand, at low temperature, i.e. when $T \to 0$, $C_V \propto (1/T^2)e^{-\hbar\omega/k_BT}$; that is, it decreases exponentially.

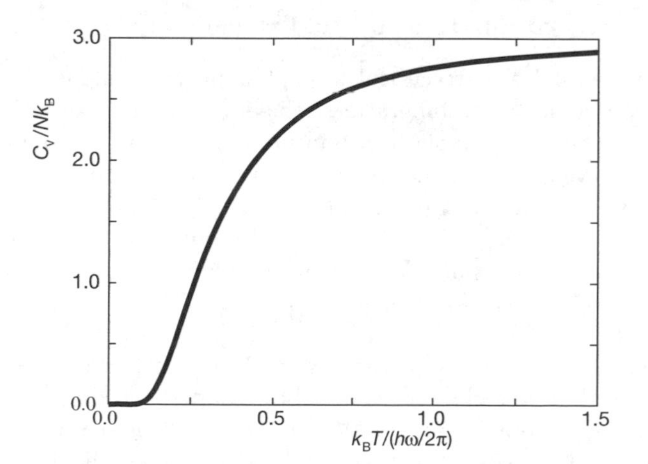

Fig. 5.2. Heat capacity due to lattice vibrations, calculated from the Einstein model. C_V/Nk_B is plotted as a function of $k_BT/\hbar\omega$

The expression for the heat capacity is essentially the same as that for the vibration of diatomic molecules, $C^{(\mathrm{V})}$. The difference is that here an atom can move in any direction, whereas only vibration in the x-direction is possible in the case of a diatomic molecule. The average energy of each atom is given by the Bose distribution as before:

$$\langle E \rangle = \left(\langle n_x \rangle + \langle n_y \rangle + \langle n_z \rangle + \frac{3}{2} \right) \hbar\omega \,, \tag{5.5}$$

where

$$\langle n_x \rangle = \langle n_y \rangle = \langle n_z \rangle = \frac{1}{\mathrm{e}^{\beta\hbar\omega} - 1} . \tag{5.6}$$

One quantitative difference between diatomic molecules and solids is that whereas the vibrations of diatomic gases are dead at room temperature, experimental evidence tells us that the vibrations of atoms in solids are in the high-temperature regime $\hbar\omega \ll k_{\mathrm{B}}T$. This difference arises from two factors. One is a difference in the "spring constant". The bond between the atoms in a diatomic molecule is a covalent bond, which is a strong chemical bond. On the other hand, the bonds between the atoms in a solid are rather weak. In the case of metals, these are metallic bonds, mediated by the conduction electrons. That these bonds are weak is reflected in the fact that metals are easily scratched. The other factor is the difference in the mass of the atoms. The atoms in the molecules of oxygen and nitrogen gas are rather light. In contrast, atoms such as gold, silver, or iron are heavy. Considering these differences, it is easy to understand why diamond has a small heat capacity at room temperature; the reason is discussed in Sect. 5.2.4 after we have studied the Debye model.

5.2 Heat Capacity of a Solid II – Debye Model

5.2.1 Collective Oscillations of the Lattice and the Internal Energy

The Einstein model was successful in explaining the Dulong–Petit law, but failed to explain the low-temperature behavior. The heat capacity decreases exponentially according to the Einstein model, much faster than the T^3 dependence observed experimentally. The failure arises from the fact that the vibrations of atoms are not independent. The atoms move as coupled oscillators, and the frequency of oscillation can be quite low when nearby atoms move collectively. The collective oscillations of a solid behave as waves, as shown in Fig. 5.3. This aspect of the oscillations is taken into account in the Debye model.

The coupled oscillations of the crystal lattice are described by normal modes; any oscillation can be represented by a superposition of normal modes. A normal mode here is a plane wave, characterized by a wave vector $\boldsymbol{k}$ and the direction of the displacement. A mode in which the displacement is parallel to the wave vector is called a *longitudinal* mode. A mode in which the displacement is perpendicular to the wave vector is called a *transverse* mode.

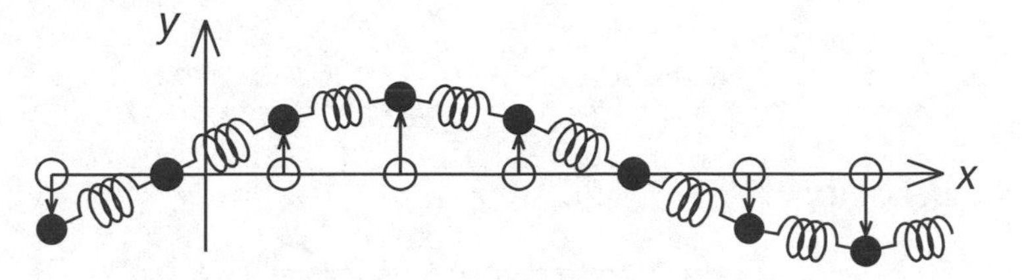

Fig. 5.3. Coupled oscillations of atoms. Here, one row of atoms aligned in the x-direction is shown. The atoms (*solid circles*) are displaced along the y-axis from the lattice points (*open circles*). In addition to the displacement in the y-direction shown here, displacements in the x- and z-directions exist

To obtain a better understanding, let us consider a one-dimensional crystal of length L first. We take the x-axis along the crystal. The atoms move in the x-direction in a longitudinal mode. There are also two sets of transverse modes, since the atoms can be displaced either in the y-direction or in the z-direction. If we assume that the atoms at both ends of the crystal are fixed, standing waves play the role of normal modes. Just as in the case of the wave function of a molecule in a box, the allowed wavelengths λ of the standing waves are given by

$$\frac{\lambda}{2}n = L, \quad n = 1, 2, 3, \cdots, N\,, \tag{5.7}$$

where $N = L/a$ is the number of atoms and a is the spacing between the atoms, or lattice constant. The allowed wave numbers are

$$k_n = \frac{2\pi}{\lambda} = \frac{\pi}{L}n\,. \tag{5.8}$$

In this case, the number of normal modes is finite. There is a shortest wavelength, as shown in Fig. 5.4. Therefore, there are N different wave numbers for each direction of displacement. Since there is one set of longitudinal modes and two sets of transverse modes, the total number of normal modes is $3N$, which coincides with the number of degrees of freedom of the N atoms.

In a normal mode, every atom oscillates at the same frequency, by definition. The angular frequency and the wave number are related:

$$\omega_{\mathrm{t}}(k) = 2\frac{v_{\mathrm{t}}}{a}\sin\frac{ka}{2} \simeq v_{\mathrm{t}}k \qquad \text{(transverse wave)}\,, \tag{5.9}$$

$$\omega_{\mathrm{l}}(k) = 2\frac{v_{\mathrm{l}}}{a}\sin\frac{ka}{2} \simeq v_{\mathrm{l}}k \qquad \text{(longitudinal wave)}\,. \tag{5.10}$$

Here v_{t} and v_{l} are the velocities (sound velocities) of the transverse and longitudinal waves, respectively, in the solid.[3] These relations between the frequency and wave number are called *dispersion relations*. Although the dis-

[3] A long-wavelength oscillation in a crystal lattice is nothing but a sound wave.

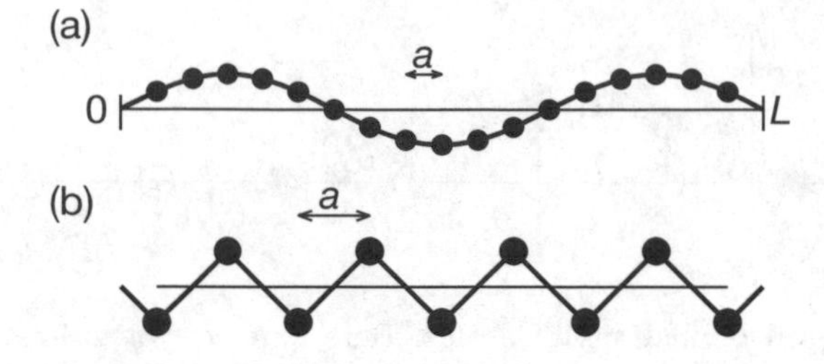

Fig. 5.4. (**a**) An example of a standing wave in a one-dimensional crystal of length L. The spacing between atoms, i.e. the lattice constant, is a. (**b**) The displacement of shortest wavelength, i.e. the standing wave with the largest wave number

persion relations of a crystal lattice are not linear, we adopt a linear approximation to the right-hand side of these equations, and use this linear dispersion in our subsequent calculations. This approximation is called the *Debye approximation.*

For a real three-dimensional crystal, the wave number is replaced by the wave vector $\boldsymbol{k}$, where

$$\boldsymbol{k} = \left(\frac{\pi}{L_x} n_x, \frac{\pi}{L_y} n_y, \frac{\pi}{L_z} n_z \right) . \tag{5.11}$$

The allowed values of $\boldsymbol{k}$ form a lattice in the wave vector space, shown in Fig. 5.5. The angular frequencies are given as follows in the Debye approximation:

$$\omega_{\mathrm{t}}(\boldsymbol{k}) = v_{\mathrm{t}} |\boldsymbol{k}| \, , \tag{5.12}$$

$$\omega_{\mathrm{l}}(\boldsymbol{k}) = v_{\mathrm{l}} |\boldsymbol{k}| \, . \tag{5.13}$$

When we treat the oscillations quantum mechanically, a normal mode of frequency ω behaves similarly to a harmonic oscillator of the same frequency. Thus, the energy of a normal mode is quantized:

$$E_{\alpha,n}(\boldsymbol{k}) = \left(n + \frac{1}{2} \right) \hbar \omega_\alpha (\boldsymbol{k}) \, , \tag{5.14}$$

where n is the quantum number of the oscillation, and $\alpha(= \mathrm{l}$ or $\mathrm{t})$ indicates the direction of the oscillation. Since the expression for the energy is the same as that for a harmonic oscillator, the thermal average of this energy is given by the same equation as before: namely, we can replace n in the right-hand side of (5.14) by $\langle n \rangle$. The total internal energy is obtained by summing this thermal average $\langle E_\alpha(\boldsymbol{k}) \rangle$ over α and $\boldsymbol{k}$:

$$\begin{aligned} U &= \sum_{\boldsymbol{k}} \sum_{\alpha} \left(\langle n \rangle + \frac{1}{2} \right) \hbar \omega_\alpha (\boldsymbol{k}) \\ &= \sum_{\boldsymbol{k}} \sum_{\alpha} \left[\frac{\hbar \omega_\alpha (k)}{\exp\left[\beta \hbar \omega_\alpha (k) \right] - 1} + \frac{1}{2} \hbar \omega_\alpha (k) \right] . \end{aligned} \tag{5.15}$$

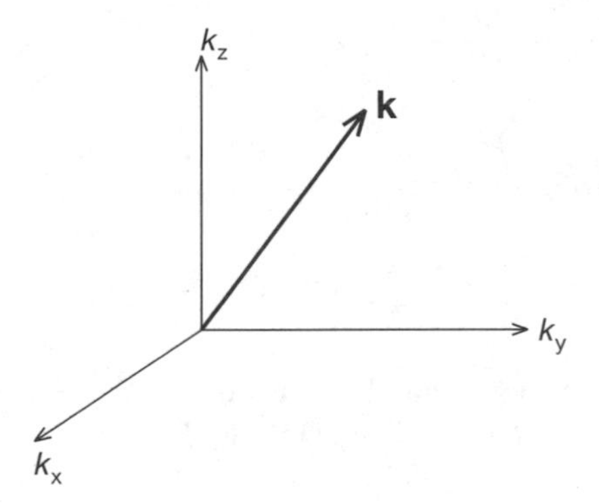

Fig. 5.5. Wave vector space. The allowed values of $\boldsymbol{k} = (\pi n_x/L_x,\ \pi n_y/L_y,\ \pi n_z/L_z n_z)$ form a lattice in this space

The final term in the right-hand side is the contribution from the zero-point oscillation, which is independent of the temperature.

To evaluate this equation, we perform the summation over $\boldsymbol{k}$ as an integral in wave vector space. In the one-dimensional case $k_n = (\pi/L)n$, there is one point corresponding to an allowed wave number for each interval $\Delta k = \pi/L$. In the three-dimensional case, there is one point corresponding to an allowed wave vector in each volume $\pi^3/L_x L_y L_z$. Therefore, there are $\mathrm{d}k_x\,\mathrm{d}k_y\,\mathrm{d}k_z/(\pi^3/V)$ allowed wave vectors in a volume of $\mathrm{d}k_x\,\mathrm{d}k_y\,\mathrm{d}k_z$, and so U is given in integral form by

$$
\begin{aligned}
U &= \sum_\alpha \int_0^{\pi/a} \mathrm{d}k_x \int_0^{\pi/a} \mathrm{d}k_y \int_0^{\pi/a} \mathrm{d}k_z \frac{V}{\pi^3} \frac{\hbar\omega_\alpha(k)}{\exp\left[\beta\hbar\omega_\alpha(k)\right] - 1} + \text{const.} \\
&= \frac{V}{8\pi^3} \sum_\alpha \int_{-\pi/a}^{\pi/a} \mathrm{d}k_x \int_{-\pi/a}^{\pi/a} \mathrm{d}k_y \int_{-\pi/a}^{\pi/a} \mathrm{d}k_z \frac{\hbar\omega_\alpha(k)}{\exp\left[\beta\hbar\omega_\alpha(k)\right] - 1} + \text{const.}
\end{aligned}
\tag{5.16}
$$

In the second line of this equation, the integrals for each direction have been extended to negative wave vector components, and the expression has been divided by $2^3 = 8$. This is allowed because the frequency is a function of the absolute value of $\boldsymbol{k}$. The last term is the contribution from the zero-point oscillation, which is independent of temperature. On the basis of this expression for the internal energy, we shall now investigate the temperature dependence of the internal energy and the heat capacity.

5.2.2 Heat Capacity at High Temperature

The high-temperature region is defined as that in which $\beta\hbar\omega_\alpha(k) \ll 1$ is satisfied for all of the normal modes. In this case the following approximation can be used:

$$
\exp[\beta\hbar\omega_\alpha(k)] \simeq 1 + \beta\hbar\omega_\alpha(k) = 1 + \frac{\hbar\omega_\alpha(k)}{k_{\mathrm{B}}T}\,. \tag{5.17}
$$

Then, apart from a constant term, we obtain

$$U = \frac{V}{8\pi^3} \sum_\alpha \int_{-\pi/a}^{\pi/a} \mathrm{d}k_x \int_{-\pi/a}^{\pi/a} \mathrm{d}k_y \int_{-\pi/a}^{\pi/a} \mathrm{d}k_z \, k_\mathrm{B}T$$

$$= \frac{V}{8\pi^3} \times 3 \times \left(\frac{2\pi}{a}\right)^3 k_\mathrm{B}T = 3\left(\frac{V}{a^3}\right) k_\mathrm{B}T = 3Nk_\mathrm{B}T\,. \qquad (5.18)$$

Here we have used the fact that V/a^3 is the total number of atoms N. The equipartition law is satisfied, and the Dulong–Petit law is obeyed by the molar heat.

5.2.3 Heat Capacity at Low Temperature

In the Debye model, the frequencies of the normal modes are distributed from $v_\alpha \pi/L$ to the highest frequency, of the order of $v_\mathrm{l}\pi/a \equiv \omega_\mathrm{max}$. We define the low-temperature region such that the thermal energy $k_\mathrm{B}T$ is much smaller than the energy unit for the highest frequency, i.e. $\beta\hbar\omega_\mathrm{max} \gg 1$. In this temperature region $\exp[\beta\hbar\omega_\mathrm{max}]$ is huge, and so the integrand in (5.16) vanishes beyond the boundary of the integral. Therefore, we can extend the integral to the whole wave vector space. Then U can be calculated analytically, except for a constant term:

$$U \simeq \frac{V}{8\pi^3} \sum_\alpha \int_{-\infty}^{\infty} \mathrm{d}k_x \int_{-\infty}^{\infty} \mathrm{d}k_y \int_{-\infty}^{\infty} \mathrm{d}k_z \frac{\hbar v_\alpha k}{\exp(\beta\hbar v_\alpha k) - 1}$$

$$= \frac{V}{8\pi^3} \sum_\alpha \int_0^\infty 4\pi k^2 \, \mathrm{d}k \frac{\hbar v_\alpha k}{\exp(\beta\hbar v_\alpha k) - 1}$$

$$= \frac{V}{2\pi^2} \frac{1}{\beta^4} \sum_\alpha \frac{1}{(\hbar v_\alpha)^3} \int_0^\infty \mathrm{d}x \frac{x^3}{\mathrm{e}^x - 1}\,. \qquad (5.19)$$

In the final line, $\beta\hbar v_\alpha k \equiv x$ has been used. The final integral with respect to x gives $\pi^4/15$. This result shows that $U \propto \beta^{-4} \propto T^4$, and hence $C_\mathrm{V} \propto T^3$.

5.2.4 Heat Capacity at Intermediate Temperature

In this region, we need to perform a numerical integration of (5.16). If we take the velocities of the longitudinal and transverse waves to be the same for simplicity, and define T_D, which is called the Debye temperature, by $k_\mathrm{B}T_\mathrm{D} \equiv \hbar\omega_\mathrm{max}$, the heat capacity can be expressed as a function of T/T_D. The result of a numerical calculation is shown in Fig. 5.6. It is known that this temperature dependence reproduces the behavior of the actual heat capacities of solids. The heat capacity starts to decrease below T_D. The values of the Debye temperature for several solids are listed in Table 5.2. We remark that $T_{\rm D}$ of diamond is exceptionally high. This is the reason that diamond does not obey the Dulong–Petit law at room temperature.

This high Debye temperature can be understood from the well-known properties of diamond. The average frequency of the normal modes is the same as the frequency in the Einstein model, and is given by $\omega = \sqrt{k/m}$, where k is the spring constant and m is the mass of an atom. Diamond is the hardest solid known, which means that the bonds (covalent) between the atoms are strong, and so k must be large. Furthermore, the carbon atom is one of the lighter atoms. Therefore, it is natural that diamond has a high Debye temperature. Lead has properties opposite to those of diamond. It is soft, and the lead atom is one of the heaviest atoms. Therefore, the Debye temperature of lead is low. We can estimate the "spring constant" k from the Debye temperature and the mass of an atom, and can compare the strengths of various kinds of bonds. This is left as an exercise for readers at the end of this chapter.

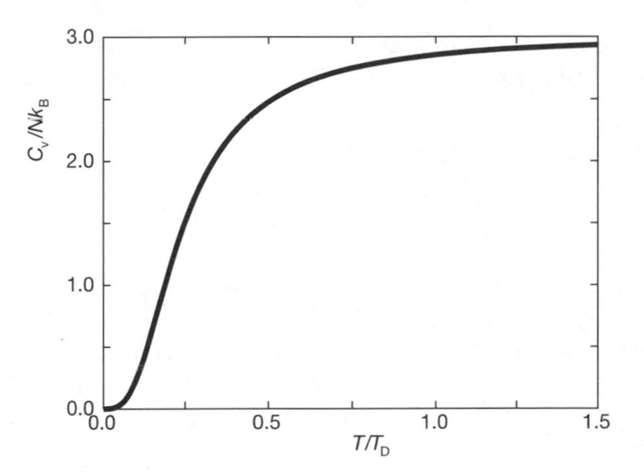

Fig. 5.6. Heat capacity calculated from the Debye model. The *horizontal axis* shows $T/T_{\rm D}$, where the Debye temperature $T_{\rm D}$ is defined by $k_{\rm B}T_{\rm D} \equiv \hbar\omega_{\rm max}$

5.2.5 Physical Explanation for the Temperature Dependence

The Dulong–Petit law, obeyed at high temperature, is easy to understand. In this case $k_{\rm B}T \gg \hbar\omega_{\rm max}$, and so the equipartition law is obeyed for all normal modes, as in the Einstein model at high temperature. The behavior at low temperature can be understood as follows. From our experience with the diatomic molecule and the Einstein model, we know that the contribution to the heat capacity from modes with frequencies ω that satisfy $\hbar\omega > k_{\rm B}T$ will be negligibly small. For these modes, $\langle n \rangle \sim 0$; these modes are dead, inert modes. Now, we can divide the modes roughly into two parts at a frequency $\omega_{\rm T} \equiv$

Table 5.2. Debye temperatures of typical solids

Material	T_{D}	Remarks
Lead (Pb)	105 K	Heavy, soft metal
Gold (Au)	165 K	Heavy, soft metal
Sodium chloride (NaCl)	321 K	Ordinary ionic crystal
Aluminum (Al)	428 K	Ordinary metal
Iron (Fe)	467 K	Ordinary metal
Diamond (C)	2230 K	Light, hard crystal

$k_{\mathrm{B}}T/\hbar$. We neglect the contribution to the heat capacity from the modes with $\omega > \omega_{\mathrm{T}}$. On the other hand, we assume that the equipartition law is obeyed for the active modes, where $\omega < \omega_{\mathrm{T}}$. The number of active modes is given by the number of points representing $\boldsymbol{k}$ values in wave vector space in a sphere of radius $k_{\mathrm{T}} = \omega_{\mathrm{T}}/v_{\alpha}$. This number is proportional to the volume of the sphere, which is proportional to $k_{\mathrm{T}}^3 \propto T^3$. The internal energy is this number of active modes times $k_{\mathrm{B}}T$, and so is proportional to T^4. This is the reason why U in (5.19) is proportional to T^4, and the heat capacity is proportional to T^3.

5.3 Black-Body Radiation

5.3.1 Wien's Law and Stefan's Law

Iron heated in a blast furnace glows red at lower temperatures, and white at higher temperatures. The temperature of the iron determines how it glows, and so can be measured by analysis of the spectrum. The relation between the spectrum and the temperature does not depend on the material at high temperature. In this section, we investigate this relationship theoretically. The electromagnetic energy emitted from a sample of matter at temperature T between frequencies ν and $\nu + \mathrm{d}\nu$ is written as $K_\nu(T)\,\mathrm{d}\nu$. The behavior of this function is shown in Fig. 5.7.

By the end of the nineteenth century, two laws were known:

- *Wien's displacement law* (1893). From thermodynamic considerations, Wien concluded that the frequency dependence of the radiated energy must have the form $K_\nu(T) = \nu^3 F(\nu/T)$. From this functional form, we can conclude that the product of the wavelength λ_m at which the intensity of the radiated energy reaches its maximum and the temperature is a constant, i.e. $\lambda_m T = \text{constant}$.
- *Stefan's law* (1879). From experiments, Stefan showed that the total energy of the radiation in a cavity was proportional to T^4. Boltzmann proved this result in 1884 using thermodynamics. Therefore, this law is also known as the Stefan–Boltzmann law.

In this section, we calculate $K_\nu(T)$, and confirm these laws.

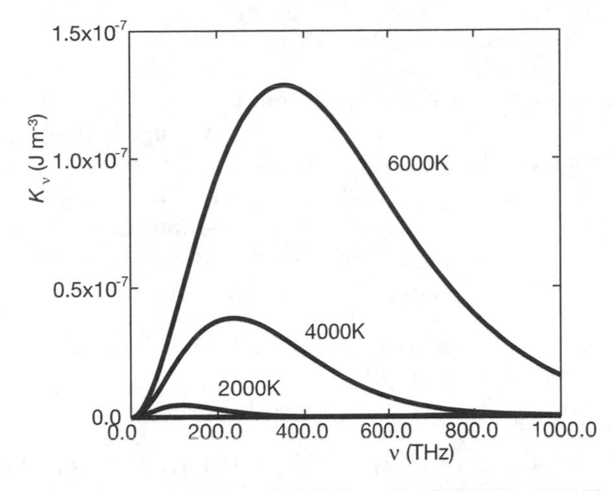

Fig. 5.7. Frequency dependence of $K_\nu(T)$ at $T = 2000\,\mathrm{K}$, $4000\,\mathrm{K}$, and $6000\,\mathrm{K}$

Every sample of matter has its intrinsic color, which affects $K_\nu(T)$, especially at low temperatures. To remove the effect of this color, we first define a *black body*. This is a sample of matter that absorbs all electromagnetic radiation incident on it. Therefore, there is no reflected light, and the light emitted from a black body originates from the fact that it has a temperature T. Such a black body can be realized by cutting a tiny hole in the wall of a cavity. Light going into the cavity is reflected by the inside wall many times before it comes out of the hole again. At each reflection, part of the light is absorbed, and so when it comes out of the hole, it is substantially weakened. In fact, when we look into the hole, it is dark inside, and we see no light. A small window in a blast furnace is such a black body; when the temperature of the furnace becomes sufficiently high, it begins to emit light.

5.3.2 Energy of Radiation in a Cavity

Light is emitted from a cavity because the space in the cavity is filled with electromagnetic radiation that is in thermal equilibrium with the wall, at a temperature T. We consider a rectangular cavity whose size is $L_x \times L_y \times L_z$. If we assume a boundary condition such that standing waves are allowed, the allowed wave vectors of the radiation are

$$\boldsymbol{k} = (k_x, k_y, k_z) = \left(\frac{\pi}{L_x}n_x, \frac{\pi}{L_y}n_y, \frac{\pi}{L_z}n_z\right), \quad n_x, n_y, n_z > 0\,. \tag{5.20}$$

We can also use a periodic boundary condition, which gives the following condition for the wave vectors:

$$\boldsymbol{k} = (k_x, k_y, k_z) = \left(\frac{2\pi}{L_x}n_x, \frac{2\pi}{L_y}n_y, \frac{2\pi}{L_z}n_z\right), \tag{5.21}$$

where $-\infty < n_x, n_y, n_z < \infty$. The angular frequency of the radiation, ω, is given by $\omega(\boldsymbol{k}) = c|\boldsymbol{k}|$, where c is the velocity of light. The following calculation does not depend on the boundary condition, and so we can use either of these conditions; the number of modes in the frequency range between ω and $\omega+\Delta\omega$ is independent of the boundary condition. We summarize these boundary conditions in Appendix F. We notice that the situation is quite similar to that for the Debye model described in the previous section. A plane wave with a wave vector $\boldsymbol{k}$ is a normal mode of the electromagnetic radiation, and behaves as a harmonic oscillator. The main differences are

- electromagnetic radiation has no longitudinal modes;
- there is no upper bound on $|\boldsymbol{k}|$;
- the velocity is not the velocity of sound but that of light.

Apart from these differences, we can apply statistical mechanics just as before. There are two polarization directions for transverse modes, and so the internal energy of the radiation is

$$\begin{aligned} U &= \frac{V}{8\pi^3} \times 2 \int_{-\infty}^{\infty} \mathrm{d}k_x \int_{-\infty}^{\infty} \mathrm{d}k_y \int_{-\infty}^{\infty} \mathrm{d}k_z \frac{\hbar ck}{\mathrm{e}^{\beta\hbar ck} - 1} + \text{const.} \\ &= \frac{V}{4\pi^3} \int_0^{\infty} \mathrm{d}k \, 4\pi k^2 \frac{\hbar ck}{\mathrm{e}^{\beta\hbar ck} - 1} + \text{const.} \\ &= \frac{V}{\pi^2} \frac{(k_{\mathrm{B}}T)^4}{(\hbar c)^3} \int_0^{\infty} \mathrm{d}k \frac{k^3}{\mathrm{e}^k - 1} + \text{const.} = V \frac{\pi^2}{15} \frac{(k_{\mathrm{B}}T)^4}{(\hbar c)^3} + \text{const.} \\ &= V \frac{8\pi^5}{15} \frac{(k_{\mathrm{B}}T)^4}{h^3 c^3} + \text{const.} \end{aligned} \tag{5.22}$$

Here, the constant is the contribution from the zero-point energy. This result agrees with Stefan's law. Since there is no maximum to the frequency, radiation in a cavity is always in the low-temperature regime: $\infty = \hbar\omega_{\max} \gg k_{\mathrm{B}}T$. Stefan's law and the T^3 dependence of the low-temperature heat capacity of a solid are closely related.

5.3.3 Spectrum of Light Emitted from a Hole

The energy of the radiation of wave vector $\boldsymbol{k}$, apart from the zero-point energy, is $\hbar ck \langle n \rangle$, where $\langle n \rangle$ at T $(= 1/k_{\mathrm{B}}\beta)$ is given as follows:

$$\langle n \rangle = \frac{1}{\mathrm{e}^{\beta\hbar ck} - 1} . \tag{5.23}$$

If the cavity has a small hole of area S, part of this energy comes out of the hole. As shown in Fig. 5.8, the energy contained in a volume of $Sc\,\Delta t \cos\theta$ will come out during a time Δt, where θ is the angle between the normal to

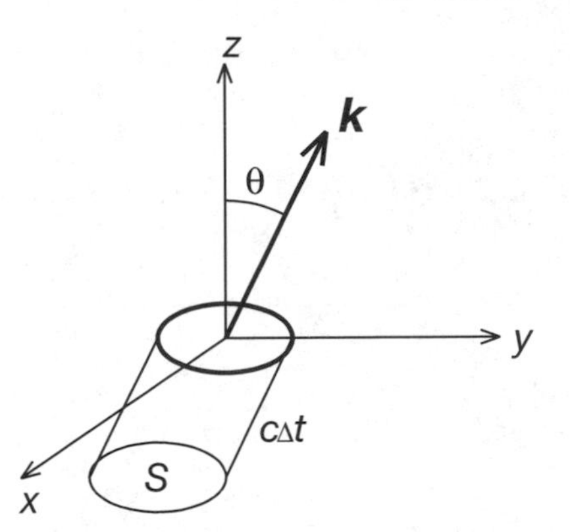

Fig. 5.8. Light with wave vector $\boldsymbol{k}$ coming out of a hole of area S. We place the hole (*thick circle*) in the xy plane, and take the z-axis perpendicular to the hole. The radiation (with wave vector $\boldsymbol{k}$) in the *canted cylinder* comes out of the *hole* during a time interval Δt

the hole (z-axis) and $\boldsymbol{k}$. Since the energy per unit volume is $\langle n \rangle \hbar c k/V$, the energy coming out of the hole is $S \cos\theta c \Delta t \langle n \rangle \hbar c k/V$, which is equivalent to $\cos\theta \langle n \rangle \hbar c^2 k/V$ per unit area per unit time.

We sum over the values of $\boldsymbol{k}$ whose energy lies between ν and $\nu + \mathrm{d}\nu$ to obtain $K_\nu(T)\,\mathrm{d}\nu$ We express $\boldsymbol{k}$ by use of polar coordinates. The number of modes in an infinitesimal volume in wave vector space defined by

$$\begin{cases} k \quad \text{to} \quad k + \mathrm{d}k \\ \theta \quad \text{to} \quad \theta + \mathrm{d}\theta \\ 0 \le \phi \le 2\pi \end{cases}$$

is $[2V/(2\pi)^3]2\pi \sin\theta\,\mathrm{d}\theta\,k^2\,\mathrm{d}k$. This number is multiplied by the emitted energy $\langle n \rangle \hbar c^2 k \cos\theta/V$ and integrated over the range $0 \le \theta \le \pi/2$. We then obtain $K_\nu(T)\,\mathrm{d}\nu$, the energy in the interval k to $k + \mathrm{d}k$:

$$\begin{aligned} K_\nu(T)\,\mathrm{d}\nu &= \int_0^{\pi/2} \mathrm{d}\theta\ \langle n \rangle \hbar c^2 k \cos\theta \frac{1}{2\pi^2} \sin\theta k^2\,\mathrm{d}k \\ &= \frac{1}{(2\pi)^2} \hbar c^2 k^3 \langle n \rangle\,\mathrm{d}k\,. \end{aligned} \tag{5.24}$$

Using the relation $k = (2\pi/c)\nu$ and $\mathrm{d}k = (2\pi/c)\,\mathrm{d}\nu$, we obtain

$$K_\nu(T) = \frac{2\pi h}{c^2} \nu^3 \frac{1}{\exp(h\nu/k_\mathrm{B}T) - 1} \equiv \nu^3 F\left(\frac{\nu}{T}\right)\,. \tag{5.25}$$

Thus we have obtained Wien's displacement law, and also obtained the unknown function as $F(x) = (2\pi h/c^2)/[\exp(hx/k_\mathrm{B}) - 1]$.

This distribution can be rewritten as the emitted energy between the wavelengths λ and $\lambda + \mathrm{d}\lambda$, $I_\lambda \, \mathrm{d}\lambda$. Using $\nu = c/\lambda$ and $\mathrm{d}\nu = -(c/\lambda^2)\,\mathrm{d}\lambda$, we obtain

$$\begin{aligned} I_\lambda(T)\,\mathrm{d}\lambda &= \frac{2\pi h}{c^2}\frac{c^3}{\lambda^3}\frac{1}{\mathrm{e}^{\beta hc/\lambda}-1}\frac{c}{\lambda^2}\,\mathrm{d}\lambda \\ &= \frac{2\pi hc^2}{\lambda^5}\frac{1}{\mathrm{e}^{\beta hc/\lambda}-1}\,\mathrm{d}\lambda\,. \end{aligned} \tag{5.26}$$

This I_λ is shown in Fig. 5.9.

The value of λ at which I_λ has its maximum, λ_m, can be written as

$$\lambda_\mathrm{m} = \frac{hc}{4.97 k_\mathrm{B} T} = \frac{2.90\times 10^{-3}}{T}\,\mathrm{m\,K^{-1}}\,. \tag{5.27}$$

When $T = 6000\,\mathrm{K}$, $\lambda_\mathrm{m} = 4.8\times 10^{-7}\,\mathrm{m} = 480\,\mathrm{nm}$. This corresponds to blue light. The surface temperature of the sun is about 6000 K, and so the spectrum of the light from the sun is peaked in the blue. Our eye-and-brain system is designed to perceive this spectrum as white light.

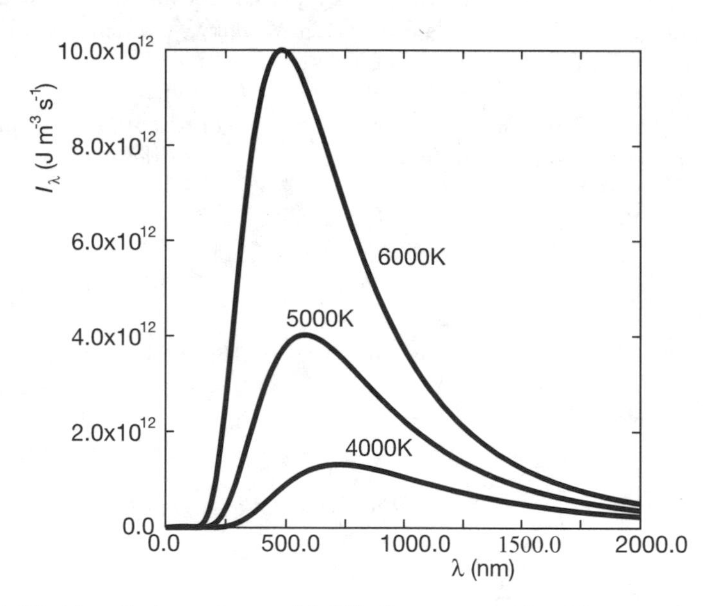

Fig. 5.9. Wavelength dependence of $I_\lambda(T)$, the energy emitted between λ and $\lambda+\mathrm{d}\lambda$ from a *hole* of unit area per second

5.3.4 The Temperature of the Universe

In 1964, Penzias and Wilson at Bell Laboratories discovered noisy microwave radiation coming from the sky while they were developing a high-sensitivity antenna for microwave communication. The radiation was peaked at $\lambda_m = 1.1\,\mathrm{mm}$. At about the same time, researchers at nearby Princeton University thought that if the universe began with a big bang and had been expanding and cooling down, it should be filled with black-body radiation. They had a plan to observe this radiation to measure the temperature of the universe. Penzias and Wilson heard of this plan, and realized that what they had considered to be noise was the black-body radiation of the universe. From the peak wavelength, the temperature of the universe was found to be 2.7 K. Penzias and Wilson were awarded a Nobel Prize in 1978.

Since its discovery, astronomers have investigated this microwave radiation, known as the cosmic microwave background radiation. The cosmic space is almost empty, but it contains stars and galaxies. Therefore, the cosmic microwave background radiation is not isotropic. In 1989 an artificial satellite called COBE was put into orbit, and it observed that there was a tiny temperature fluctuation of $\delta T/T = 10^{-5}$. In 2001, another satellite, the Wilkinson Microwave Anisotropy Probe (WMAP), was launched, and has been sending us more detailed data with higher spatial resolution. From these data, the age of the universe has been determined to be 13.7 ± 0.2 billion years. Furthermore, the data indicate that the amount of ordinary matter in the universe is only 4% of the total matter and energy, the rest being dark matter (23%) and dark energy (73%). The data also show that the inflation theory of the universe is consistent with observation [1].

Exercise 14. The Debye temperature T_D can be considered to give the average angular frequency ω of lattice vibrations, i.e. $k_\mathrm{B}T_\mathrm{D} = \hbar\omega$. The spring constant of the lattice k can be estimated from this ω using $k = m\omega^2$, where m is the mass of an atom or ion constituting the crystal. Obtain values of k for the following crystals, and compare the results with that for the oxygen molecule evaluated in Exercise 13 in Chap. 4. (1) Diamond: $T_\mathrm{D} = 2230\,\mathrm{K}$, relative atomic mass 12.01. (2) Iron: $T_\mathrm{D} = 467\,\mathrm{K}$, relative atomic mass 55.85. (3) Lead: $T_\mathrm{D} = 105\,\mathrm{K}$, relative atomic mass 207.2. (4) Sodium chloride: $T_\mathrm{D} = 321\,\mathrm{K}$, relative atomic masses 22.99 (Na) and 35.45 (Cl).

6

The Elasticity of Rubber

In this chapter, we apply the microcanonical distribution to a simple model of rubber, and explain its elasticity. It is shown that this simple model and statistical mechanics reproduce several characteristics of rubber, such as the fact that it obeys Hooke's law. The elasticity of rubber looks similar to the elasticity of a spring, but these two phenomena have different origins. The elasticity of a spring arises from the forces between metal ions, and to derive Hooke's law we need to know about quantum mechanics, the theory of electrons in metals, and so on. On the other hand, the elasticity of rubber arises from entropy, and can be understood by the use of a simple model.

6.1 Characteristics of Rubber

Rubber is distinct from ordinary solids. The special characteristics of rubber are listed below:

- It is easily deformed by weak forces.
- It can be deformed heavily: it can be elongated by two to three times its original length.
- Its Young's modulus is roughly proportional to T.
- It becomes warmer when extended rapidly, and shrinks when warmed.

Let us elaborate on these characteristics. The Young's modulus of rubber is compared in Table 6.1 with values for other typical materials to show its unusual smallness. Young's modulus E is defined as follows: when a sample of a material of length L and cross section S is pulled with a force F and elongated by ΔL, $\Delta L/L$ is proportional to F/S as long as $\Delta L/L$ is small enough. This relation is written as $F/S = E\,\Delta L/L$, where E is the Young's modulus. When a metal sample is elongated beyond its elastic limit of a few percent, it does not return to its original length after the force is removed, but a sample of rubber will return to its original shape even if it is elongated by twice its length. The Young's modulus of an ordinary material is almost temperature-independent, but it is proportional to T for rubber. The fact that rubber shrinks when warmed is opposite to the behavior of ordinary materials. You can feel how rubber warms up when extended by pulling it between your lips.

Table 6.1. Young's modulus E for typical materials

Material	Young's modulus (MPa)
Steel	2.0×10^5
Copper	1.3×10^5
Glass	0.8×10^5
Rubber	1–3

These characteristics of rubber remind us of those of ideal gases, except that "extended" in the case of rubber must be replaced by "compressed" in the case of a gas. Namely, gases have the following characteristics:

- Gases can be compressed by a weak pressure.
- It is possible to change the volume by a factor of several times.
- The bulk modulus is proportional to the temperature.
- A gas warms up when compressed adiabatically, and expands when heated.

Here, the bulk modulus K is defined by the equation $\Delta P = -K\,\Delta V/V$. This K can be calculated using $PV = nRT$:

$$\Delta P = nRT\,\Delta\left(\frac{1}{V}\right) = -nRT\frac{\Delta V}{V^2} = -\frac{P}{V}\,\Delta V\,. \tag{6.1}$$

Therefore $K = P$, and at a fixed volume V, it is proportional to T. The atmospheric pressure is about $10^3\,\mathrm{hPa} = 0.1\,\mathrm{MPa}$, and so is one order of magnitude smaller than the Young's modulus of rubber.

6.2 Model of Rubber

The characteristics of a gas can be reproduced by a simple model in which molecules move freely and collide with the walls to create a pressure. Likewise, we can construct a simple model for rubber. Rubber is a polymer; that is, it is composed of a collection of long, chain-like molecules. We consider a model in which the chains do not interact with each other, and their shape can be freely changed without force. Since the chains are independent, we shall consider only a single chain. We model a single polymer chain as being composed of rigid rods of length a connected to each other at their ends. The angle between two adjacent rods is assumed to take only one of two values: 0 or π. These two configurations are assumed to have the same energy. The model is depicted in Fig. 6.1.

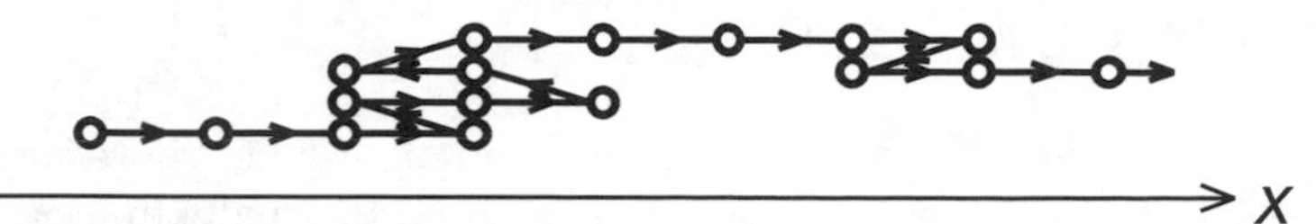

Fig. 6.1. A simplified model of rubber. A rubber molecule is modeled by a chain consisting of rigid rods. The rigid rods are shown as *arrows* with *circles* at both ends. The rods are assumed always to be parallel to the x-axis

6.3 Entropy of Rubber

We shall now calculate the entropy of this model system, and obtain a relation between force and length for the system. We assume that the total number of rods is N. Therefore, the maximum length of the chain is Na. As shown in Fig. 6.1, we treat the rods as vectors. When all the rods points in the positive x-direction, the length is at its maximum. In an ordinary state of the system, each rod can point in either the positive or the negative direction. When there are N_+ rods pointing in the positive direction and N_- $(= N - N_+)$ rods pointing in the negative direction, the total length is $x = (N_+ - N_-)a$. We calculate the entropy when the total length is x. This is given by the logarithm of the number W of microscopic states that realize this length x. This number is equal to the number of ways of choosing N_+ objects from N. Noting that

$$N_+ = \frac{Na + x}{2a} \quad \text{and } N_- = \frac{Na - x}{2a}, \tag{6.2}$$

we obtain $W(x)$:

$$W(x) = {}_N C_{N_+} = \frac{N!}{N_+!N_-!} \simeq \frac{N^N}{N_+^{N_+} N_-^{N_-}}. \tag{6.3}$$

The entropy is then given by

$$\begin{aligned} S(x) &= k_\mathrm{B} \ln W(x) \simeq k_\mathrm{B} \left[N \ln N - N_+ \ln N_+ - N_- \ln N_- \right] \\ &= k_\mathrm{B} \left[N \ln N - \left(\frac{N}{2} + \frac{x}{2a} \right) \ln \left(\frac{N}{2} + \frac{x}{2a} \right) \right. \\ &\qquad \left. - \left(\frac{N}{2} - \frac{x}{2a} \right) \ln \left(\frac{N}{2} - \frac{x}{2a} \right) \right] \\ &= k_\mathrm{B} N \left[\ln 2 - \frac{1}{2} \left(1 + \frac{x}{Na} \right) \ln \left(1 + \frac{x}{Na} \right) \right. \\ &\qquad \left. - \frac{1}{2} \left(1 - \frac{x}{Na} \right) \ln \left(1 - \frac{x}{Na} \right) \right]. \end{aligned} \tag{6.4}$$

6.4 Hooke's Law

To calculate the force needed to keep the chain at a length x, we consider the situation shown in Fig. 6.2. The system is isolated thermally from its environment, and so the total energy E is conserved. The gas in the cylinder and the sample of rubber (a chain) are in thermal equilibrium at a temperature T.[1] They are also in mechanical equilibrium: the chain is forced to have a length x by the pressure of the gas. In this equilibrium, the length of the chain and the volume of the gas are correlated, and are determined by the condition that the total entropy of the gas and rubber is maximized. We write the entropy of the gas as S_{G}, and let the cross section of the cylinder be A. The condition that the total entropy $S_{\mathrm{tot}} = S + S_{\mathrm{G}}$ is maximized with respect to x is

$$\begin{aligned} 0 &= \left(\frac{\partial S_{\mathrm{G}}}{\partial x}\right)_E + \left(\frac{\partial S}{\partial x}\right)_E = A\left(\frac{\partial S_{\mathrm{G}}}{\partial (Ax)}\right)_E + \left(\frac{\partial S}{\partial x}\right)_E \\ &= A\frac{P}{T} + \left(\frac{\partial S}{\partial x}\right)_E . \end{aligned} \tag{6.5}$$

In this equation, P is the pressure of the gas. On the other hand, the condition for mechanical equilibrium tells us that $F = PA$, where F is the tension in the chain. Therefore, the derivative of the entropy is related to the tension:

$$\left(\frac{\partial S}{\partial x}\right)_E = -\frac{F}{T} . \tag{6.6}$$

We evaluate the derivative on the left-hand side and obtain the tension:

$$F = -T\frac{\partial S}{\partial x} = \frac{k_{\mathrm{B}}T}{2a}\ln\left(\frac{1 + x/Na}{1 - x/Na}\right) \simeq \frac{k_{\mathrm{B}}T}{Na^2}x , \tag{6.7}$$

where $Na \gg x$ has been used.

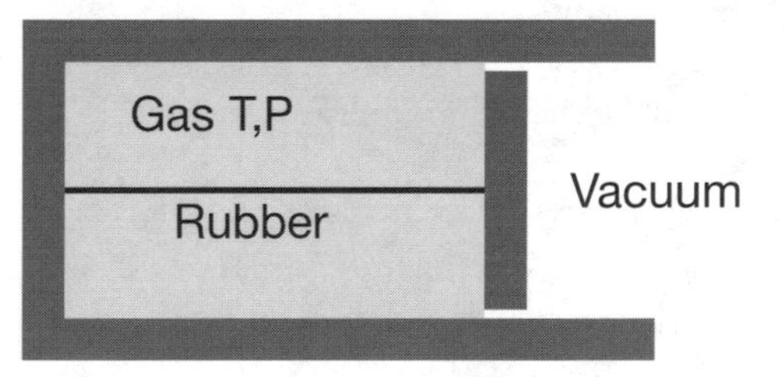

Fig. 6.2. Mechanical equilibrium between forces exerted by rubber and gas pressure. A gas at temperature T is confined in a cylinder by a force from a sample of rubber that holds the piston in place

[1] The chain in this model has no internal energy. The temperature of the gas is determined by the total energy of the system, which is equal to the energy of the gas.

This result reproduces the characteristics of rubber. The model chain obeys Hooke's law, i.e. $F \propto x$, and the elasticity constant, or Young's modulus, is proportional to the temperature T. When the chain is extended, the entropy $S(x)$ decreases, and so the excess entropy must be given to the environment. That is, the chain gives heat to its environment, which is the gas in this case. On the other hand, if heat is given to the chain, its entropy increases, and so it must shrink. In this way, we can explain the main properties of rubber with this simple model. This is an example of the virtue of statistical physics.

The present model, however, has some defects arising from oversimplification. One of them is that the total length of the chain is zero when no tension is applied. The origin of this defect is that we have neglected the thickness of the polymer chain, and have also neglected the fact that the chains are entangled three-dimensionally. To incorporate these effects is not an easy task, and can be done only approximately. The task of constructing a more realistic model and investigating it statistical-mechanically is one of the frontline fields of research in statistical physics [2, 3].

7

Magnetic Materials

We have applied the methods of statistical physics to several systems in the previous chapters. The systems that we have considered so far are simple, since they are noninteracting systems. In this chapter we investigate the Ising model as a model for ferromagnetic systems, in which interaction plays an important role. A ferromagnetic system undergoes a phase transition at high temperature to a paramagnetic phase. We discuss this transition by use of a mean-field theory. We also calculate the exact free energy for a one-dimensional system.

7.1 Origin of Permanent Magnetism

Some kinds of atoms have a magnetic moment. That is, those atoms behave as small permanent magnets. When a crystal contains such atoms, it is a realization of a system in which tiny magnetic moments are placed regularly in a crystal lattice. Under suitable conditions, these magnetic moments become aligned in a common direction. The whole crystal then behaves as a permanent magnet. When this alignment happens, the crystal is said to be *ferromagnetic*. At higher temperatures, the moments in a ferromagnetic material cease to be aligned, because of thermal fluctuations; it then becomes *paramagnetic*.

In this section, we explain what a magnetic moment of an atom is, and why these magnetic moments become aligned. First, we shall explain the magnetic moment. An atom consists of electrons and a nucleus. The nucleus consists of protons and neutrons. These particles have a spin angular momentum $\boldsymbol{S}$, which can be considered as a rotation of the particle. The magnitude of $\boldsymbol{S}$ is fixed at $\hbar/2$ for these particles. This rotation leads to a circulating electric current, and a magnetic moment $\boldsymbol{\mu}$. It is evident that this moment should be proportional to the angular velocity and the charge of the particle. For the same angular momentum, a heavier system rotates more slowly, and so we expect the magnetic moment of a nucleon to be smaller than that of an electron. The size of μ for the electron, proton, and neutron is given in Table 7.1. The reason that the neutron has a nonzero μ is that it consists of charged quarks.

Table 7.1. Magnetic moments of elementary particles. The masses of these particles are written as m_e, m_p, and m_n for the electron, proton, and neutron, respectively

Particle	Magnetic moment
Electron	$\mu_e = 9.28 \times 10^{-24}\,\mathrm{J/T} \simeq (1/2)(e\hbar/m_e)$
Proton	$\mu_p = 1.41 \times 10^{-26}\,\mathrm{J/T} = 5.58 \times e\hbar/4m_p$
Neutron	$\mu_n = 9.66 \times 10^{-27}\,\mathrm{J/T} = 4.49 \times e\hbar/4m_n$

We can see that $\mu_e \gg \mu_p, \mu_n$ as expected, and so the magnetic moment of an atom originates mostly from the electrons. Although the orbital motion of the electrons in an atom can contribute to the magnetic moment when the atom is isolated, such moments are canceled in a crystal. In the case of an iron atom, the configuration of the 26 electrons in the atom is as follows: $1s^2 2s^2 2p^6 3s^2 3p^6 3d^6 4s^2$. This means that there are two electrons in the $1s$ orbital, two electrons in the $2s$ orbital, and so on. The part $1s^2 2s^2 2p^6 3s^2 3p^6$ of the configuration is the same as the configuration of the electrons in an argon atom. Here all the allowed electronic states in the $1s$, $2s$, $2p$, $3s$, and $3p$ orbitals are filled with electrons, and it is said that *closed shells* are formed. For a closed shell, the total orbital and total spin angular momenta are zero, and there is no magnetic moment. For an iron atom, however, there are additional electrons in the $3d$ and $4s$ orbitals. The spins of the six electrons in the $3d$ orbitals are mostly aligned, and therefore they contribute to the magnetic moment of the atom. On the other hand, their orbital motion does not contribute to the magnetic moment, because each d electron is in a state that is a superposition of two counterrotating states. The two electrons in the $4s$ orbitals behave as conduction electrons in metallic iron.

The reason why the spins align can be understood as follows. Within an atom, electrons usually occupy orbitals from the lowest energy up. Owing to the Pauli principle, electrons of the same spin direction cannot occupy the same orbital, and so the two electrons in a $1s$ or $2s$ orbital have opposite spins, and their magnetic moments cancel. However, in the case of the $3d$ orbitals, there are five orbitals with energies that are almost the same. We say that these orbitals are almost *degenerate*. In this case the Coulomb interaction energy between the electrons is lower when the electrons have their spins aligned. Therefore, in an iron atom five spin-aligned electrons occupy all of the $3d$ orbitals, and the remaining electron has its spin reversed and occupies the lowest-energy $3d$ orbital, as shown in Fig. 7.1. This fact is in accordance with Hund's rules. Therefore, the six electrons in the $3d$ orbitals of iron are nearly all aligned in the same direction.

The Coulomb interaction between electrons, the Pauli principle, and the existence of the conduction electrons cause an interaction between the spins of nearby atoms. This effective interaction between spins is usually called the *exchange interaction*. The sign of the exchange interaction may be either positive or negative, depending on the material. It tends to align the spins

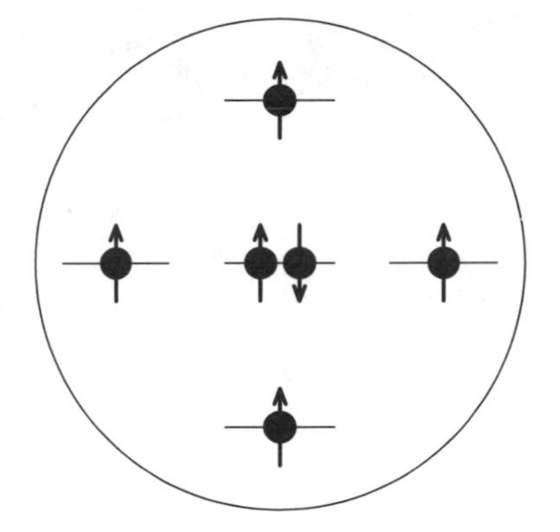

Fig. 7.1. Spin configuration of d electrons in an iron atom. In an atom, shown symbolically by the *large circle*, there are five nearly degenerate $3d$ orbitals, shown symbolically by five *horizontal lines*. These orbitals are occupied by six electrons in the case of an iron atom. The electrons are shown by *closed circles* with an *arrow* showing the direction of the spin. The first five electrons occupy these five orbitals one by one, and the spins are aligned in accordance with Hund's rules. The sixth electron must have a reversed spin to occupy one of the already occupied d orbitals, and cancels the spin of the other electron in the orbital. As a result there remain four aligned spins, and the total spin angular momentum is four times that of an electron

of nearest-neighbor atoms for some materials, such as metallic iron (Fe) and chromium dioxide (CrO_2), whereas it has the opposite effect for some other materials, such as iron(II) oxide (FeO) and cobalt(II) oxide (CoO).[1] The effect of the exchange interaction is much stronger than the dipole–dipole interaction between the magnetic moments of atoms.

7.2 Statistical Mechanics of a Free Spin System

7.2.1 Model and Entropy

We consider a model in which atoms of a single species are placed on a crystal lattice. There are N atoms per unit volume, and each atom has a nonzero total electron spin and a nonzero magnetic moment $\boldsymbol{m}$. We write the magnitude of the magnetic moment of an atom as μ. In the present form of the model, we assume that there is no interaction between the spins. Furthermore, we assume that each spin can have only one of two directions, either parallel to the z-axis

[1] Magnetite (Fe_3O_4) is also an iron oxide, and was the first permanent-magnet material that humans discovered. The three iron ions in a unit cell of the crystal are on two types of sites: one ion is on an A site and two ions are on B sites. The exchange interaction acts so as to make the spins of the ions on the A and B sites antiparallel. Since there are more ions on the B sites, a nonzero total magnetization appears in this material.

or antiparallel to it. Therefore, the magnetic moment has only a z-component, and is allowed to have only two quantized values, $m_z = \pm\mu$. This model is not as strange as it might appear, because the projection of the spin of an electron onto any direction is known from quantum mechanics to have only two possible values, $\pm\hbar/2$. We call the state of an atom that has a moment μ the "up" state and the state that has a moment $-\mu$ the "down" state. When interaction is added to this model, we obtain the Ising model, which we consider in the next section, but here we consider the noninteracting model.

The total magnetic moment of the system per unit volume is called the *magnetization*, which we write as M. When all the moments are aligned in the positive direction, the magnetization takes the value $M_{\mathrm{max}} \equiv N\mu$. This value M_{max} is called the *saturation magnetization*. The magnetization is bounded by this value:

$$-M_{\mathrm{max}} \leq M \leq M_{\mathrm{max}} \,. \tag{7.1}$$

We shall now calculate the entropy for a given value of M. We consider a system of unit volume and write the numbers of atoms in the up and down states as N_+ and N_-, respectively. The magnetization is

$$M = (N_+ - N_-)\mu \,, \tag{7.2}$$

where

$$N = N_+ + N_- \tag{7.3}$$

is the total number of atoms. That is,

$$N_\pm = \frac{1}{2\mu}\left(M_{\mathrm{max}} \pm M\right) , \tag{7.4}$$

or

$$\frac{N_\pm}{N} = \frac{1}{2}\left(1 \pm \frac{M}{M_{\mathrm{max}}}\right) . \tag{7.5}$$

Here we notice that the situation is similar to that for rubber. There are the following correspondences between the two cases:

$$\begin{aligned} N_\pm &\leftrightarrow N_\pm \,, \\ a &\leftrightarrow \mu \,, \\ x &\leftrightarrow M \,, \\ Na &\leftrightarrow M_{\mathrm{max}} \,. \end{aligned}$$

Using the notation $M/M_{\mathrm{max}} \equiv x$, we obtain the entropy per unit volume:

$$S(M) = k_{\mathrm{B}}N\left[\ln 2 - \frac{1}{2}(1+x)\ln(1+x) - \frac{1}{2}(1-x)\ln(1-x)\right] . \tag{7.6}$$

In general, the entropy is a function of the external constraints, or state variables. When we differentiate S with respect to these variables, we obtain other state variables. For example, the derivative of S with respect to the volume gives the pressure divided by the temperature, P/T. When we investigated rubber, we showed that the tension could be obtained by differentiating S

with respect to the total length x. In the present case, however, M is not an external constraint that we can fix arbitrarily. Rather, it is induced by applying magnetic field B in the z direction. Therefore, we should not expect to be able to obtain a state variable by differentiating S with respect to M. However, this model is an exception. The total energy of the system $E = -MB$ is given by the interaction between the external magnetic field B and the magnetization M. Thus,

$$\frac{\mathrm{d}S}{\mathrm{d}M} = \frac{\mathrm{d}S}{\mathrm{d}E}\frac{\mathrm{d}E}{\mathrm{d}M} = -\frac{B}{T}. \tag{7.7}$$

Differentiating (7.6) with respect to M, we obtain

$$B = \frac{k_{\mathrm{B}}T}{2\mu}\ln\left(\frac{1+x}{1-x}\right). \tag{7.8}$$

This can be rewritten as

$$x = \tanh\left(\frac{\mu B}{k_{\mathrm{B}}T}\right). \tag{7.9}$$

We must remember that this is a special result for a noninteracting system, which cannot be generalized.[2]

7.2.2 Free Energy, Magnetization, and Susceptibility

We can also treat the system by use of the canonical distribution. We assume that the system is in thermal equilibrium with a heat bath at temperature T. We also assume that the system is in a magnetic field B parallel to the z-axis. Each atom has a magnetic moment $m_z = \pm\mu$, and so it has an energy $\mp\mu B$ in the magnetic field. Thus, the partition function of an atom is

$$Z(T,B) = \mathrm{e}^{\beta\mu B} + \mathrm{e}^{-\beta\mu B}, \tag{7.10}$$

and the average moment of an atom is

$$\langle m_z \rangle = \frac{\mu\mathrm{e}^{\beta\mu B} - \mu\mathrm{e}^{-\beta\mu B}}{Z} = \mu\tanh(\beta\mu B). \tag{7.11}$$

The partition function, the free energy, and the magnetization of the total system of N atoms are

$$Z_N(T,B) = Z^N = \left(\mathrm{e}^{\beta\mu B} + \mathrm{e}^{-\beta\mu B}\right)^N = \left[2\cosh\frac{\mu B}{k_{\mathrm{B}}T}\right]^N, \tag{7.12}$$

$$F(T,B) = -Nk_{\mathrm{B}}T\ln\left[2\cosh\frac{\mu B}{k_{\mathrm{B}}T}\right], \tag{7.13}$$

[2] The function $\tanh x$ is one of the hyperbolic functions. For the definitions and properties of the hyperbolic functions, see Appendix E.

and

$$M(T,B) = N\langle m_z \rangle = N\mu \tanh\left(\frac{\mu B}{k_\mathrm{B}T}\right) . \tag{7.14}$$

The magnetic moment can also be obtained from the free energy:

$$\begin{aligned} M(T,B) &= -\left(\frac{\partial F}{\partial B}\right)_T = N\frac{1}{\beta}\frac{\partial}{\partial B}\ln Z \\ &= N\mu \tanh\left(\frac{\mu B}{k_\mathrm{B}T}\right) . \end{aligned} \tag{7.15}$$

Thus (7.8) and (7.9) have been reproduced.

The entropy as a function of T and B is

$$\begin{aligned} S(T,B) &= -\left(\frac{\partial F}{\partial T}\right)_B \\ &= k_\mathrm{B}N\left\{\ln\left[2\cosh\left(\frac{\mu B}{k_\mathrm{B}T}\right)\right] - \frac{\mu B}{k_\mathrm{B}T}\tanh\left(\frac{\mu B}{k_\mathrm{B}T}\right)\right\} . \end{aligned} \tag{7.16}$$

As $T \to \infty$ or $B \to 0$, this entropy tends to $k_\mathrm{B}N\ln 2$. The temperature dependence is shown in Fig. 7.2. If we eliminate B from this equation using (7.8), we return to the expression for the entropy (7.6). Thus, the expression for the total differential of the free energy,

$$\mathrm{d}F(T,B) = -S(T,B)\,\mathrm{d}T - M(T,B)\,\mathrm{d}B , \tag{7.17}$$

has been confirmed.

The magnetic-field dependence of $M(T,B)$ is shown in Fig. 7.3. This figure shows that a magnetization is induced by the magnetic field. When B is small,

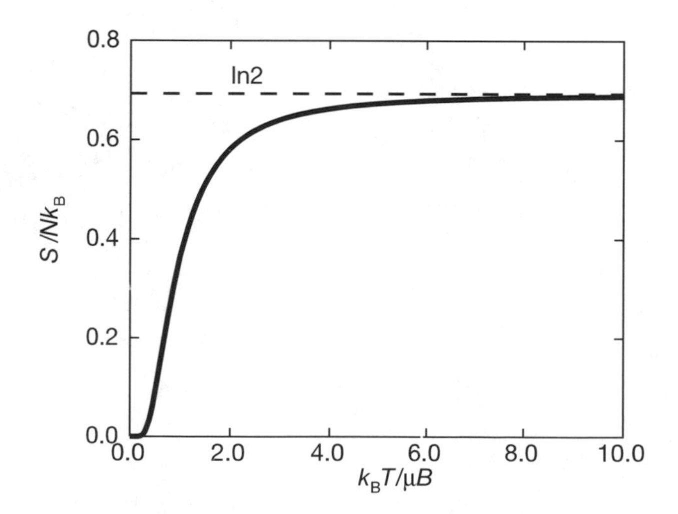

Fig. 7.2. Temperature dependence of the entropy at nonzero magnetic field B

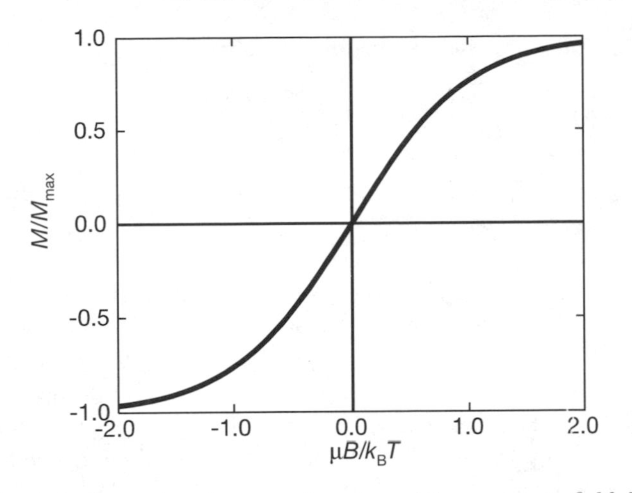

Fig. 7.3. Magnetization as a function of the magnetic field B

M is proportional to B. The proportionality constant defines the magnetic susceptibility, written as $\chi(T)$. Usually, the magnetic field H is used instead of B to express the proportionality: $\chi = (\partial M/\partial H)_{H=0}$. In the present model, in which interactions between atoms are neglected, $B = \mu_0 H$, where $\mu_0 = 4\pi \times 10^{-7}\,\mathrm{H\,m^{-1}}$ is the permeability of free space.[3] Thus,

$$\chi(T) \equiv \mu_0 \left.\frac{\partial M}{\partial B}\right|_{B=0} = \mu_0 \frac{\mu}{k_\mathrm{B}T} M_{\max} = \mu_0 \frac{N\mu^2}{k_\mathrm{B}T} \propto \frac{1}{T}\,. \tag{7.18}$$

The susceptibility is dimensionless and inversely proportional to the temperature. This temperature dependence is known as *Curie's law*, and is obeyed by paramagnetic materials. From the coefficient, we can obtain experimentally the magnitude of the magnetic moment μ of an atom.

7.2.3 Internal Energy and Heat Capacity

Next, we shall calculate the average energy and the heat capacity. The average energy $\tilde{E} \equiv \langle E \rangle$ in a magnetic field can be obtained by means of a Legendre transformation of F:[4]

$$\tilde{E}(S, B) = F + TS = -N\mu B \tanh\left(\frac{\mu B}{k_\mathrm{B}T}\right) = -MB\,. \tag{7.19}$$

[3] The correct relation between the magnetic fields H and B is $B = \mu_0(H+M)$. The contribution from the magnetization M is due to the magnetic dipole interaction. In the present model this interaction has been neglected. Thus $B = \mu_0 H$ has been used here. For real materials, M is usually much smaller than H, except when the material is ferromagnetic.

[4] Since the energy here contains the effect of the external magnetic field, we cannot call $\tilde{E}$ the internal energy.

This result can also be obtained from the general relation

$$\tilde{E}(S,B) = -\frac{\partial \ln Z_N}{\partial \beta}\,. \tag{7.20}$$

This result, $\tilde{E} = -MB$, is as it should be. From $\tilde{E}$, we obtain the heat capacity at constant magnetic field:

$$C = \left(\frac{\partial \tilde{E}}{\partial T}\right)_B = Nk_{\mathrm{B}}\left(\frac{\mu B}{k_{\mathrm{B}}T}\right)^2 \operatorname{sech}^2\left(\frac{\mu B}{k_{\mathrm{B}}T}\right)\,. \tag{7.21}$$

The temperature dependence of the average energy and of the heat capacity is shown in Fig. 7.4. The heat capacity is peaked around $k_{\mathrm{B}}T \simeq \mu B$, and tends to zero at high temperature. This behavior is quite different from that of the heat capacity of a solid.

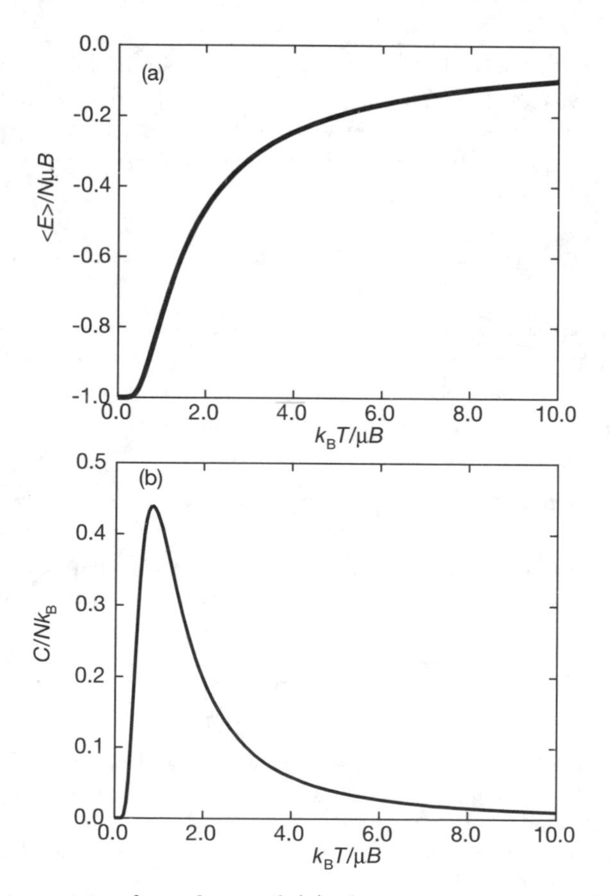

Fig. 7.4. Temperature dependence of (**a**) the average energy and (**b**) the heat capacity of a free spin system

The average energy $\tilde{E}(S,B)$ is a function of S and B, and the total differential is

$$\mathrm{d}\tilde{E}(S,B) = \mathrm{d}F + T\,\mathrm{d}S + S\,\mathrm{d}T = T\,\mathrm{d}S - M\,\mathrm{d}B\,. \tag{7.22}$$

On the other hand, the true internal energy is obtained by removing the contribution of the external field B:

$$U(S,M) = \tilde{E}(S,B) + MB = 0\,. \tag{7.23}$$

This can also be considered as a Legendre transformation, and so this U is a function of S and M. As the energy in this model comes solely from the interaction with the magnetic field, this result $U = 0$ is reasonable. From the differential of the internal energy, which is also zero, we obtain the following relation:

$$\mathrm{d}U(S,M) = T(S,M)\,\mathrm{d}S + B(S,M)\,\mathrm{d}M = 0\,. \tag{7.24}$$

In the present system $S(M)$ is a function of M only, and so we obtain (7.7) again from this relation between differentials:

$$\frac{\mathrm{d}S}{\mathrm{d}M} = -\frac{B}{T}\,.$$

7.3 Ising Model – Mean-Field Approximation

In the previous section, we neglected the interaction of a spin with surrounding spins, and obtained the result that the magnetization vanishes as the magnetic field goes to zero. However, when we take interactions between spins into account, a ferromagnetic phase becomes possible. In this case the magnetization remains nonzero even without a magnetic field. We shall investigate how a nonzero magnetization emerges by using an approximation called the *mean-field approximation*. This approximation gives a qualitatively correct description of the phenomenon in a three-dimensional isotropic system. It is known to be a better approximation in a fictitious four-dimensional system.[5]

7.3.1 Links

We consider the same model as in the previous section except that here we take interaction between nearest-neighbor spins into account. We draw fictitious

[5] On the other hand, for lower-dimensional systems, the mean-field approximation becomes worse. However, for some one- and two-dimensional model systems, an exact treatment is possible. We shall study exact solutions of the one-dimensional and two-dimensional Ising models in Sects. 7.4 and 9.3, respectively.

lines between the members of every nearest-neighbor pair, and call these lines links. For example, when atoms with spins are placed on a simple cubic lattice, each atom has six nearest neighbors, and six links are attached to the central spin. As shown in Fig. 7.5, there are two categories of links. One category is that of parallel links, in which the two spins at the ends of the link are parallel to each other. The other is that of antiparallel links, in which the spins point in opposite directions. Owing to the interaction, the energy of a link depends on whether it is parallel or antiparallel. In the model considered here, a parallel link has an energy $-J$, and an antiparallel link has an energy $+J$. If there are N spins and each spin has z nearest neighbors,[6] the total number of links is $zN/2$.

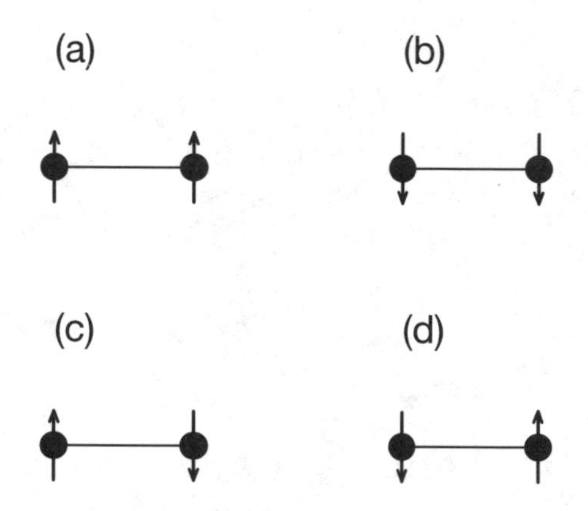

Fig. 7.5. Two nearest-neighbor atoms (*solid circles*) and a link between them. The spin of an atom is shown by an *arrow*. (**a**), (**b**) Parallel links in which the spins at the ends of the link are parallel to each other. (**c**), (**d**) Antiparallel links

A microscopic state is described by arrangement of up and down spins. As in the previous section, we use N_+ and N_- to represent the numbers of up spins and down spins, respectively. We also define N_{++} as the number of links in which the two spins at both ends are up. We define N_{+-}, N_{-+}, and N_{--} similarly. We then have the following relations for the total number of spins N, the total number of links $zN/2$, the magnetization M, and the interaction energy $E_{\rm i}$:

$$
\begin{aligned}
N &= N_+ + N_- \,, \\
\frac{1}{2}zN &= N_{++} + N_{+-} + N_{-+} + N_{--} \,, \\
M &= (N_+ - N_-)\,\mu \,, \\
E_{\rm i} &= -J\,(N_{++} + N_{--} - N_{+-} - N_{-+}) \,. \qquad (7.25)
\end{aligned}
$$

[6] For a simple cubic lattice, $z = 6$.

7.3.2 Mean-Field Approximation

To proceed further, we assume that the N_+ up spins and N_- down spins are distributed randomly. This is the mean-field approximation. The probability of finding an up-spin atom is then N_+/N, and the probability of finding a ++ link is $(N_+/N)^2$; the average numbers for the four kinds of links are

$$\langle N_{++}\rangle = \frac{1}{2}zN\left(\frac{N_+}{N}\right)^2 = \frac{z}{8}N\left(1+x\right)^2 , \tag{7.26}$$

$$\langle N_{--}\rangle = \frac{z}{8}N\left(1-x\right)^2 , \tag{7.27}$$

$$\langle N_{+-}\rangle = \langle N_{-+}\rangle = \frac{1}{2}zN\left(\frac{N_+}{N}\right)\left(\frac{N_-}{N}\right) = \frac{z}{8}\left(1-x^2\right) , \tag{7.28}$$

where $x \equiv M/M_{\max}$ as before. The average of the interaction energy is

$$\langle E_{\mathrm{i}}\rangle = -\frac{z}{2}JNx^2 . \tag{7.29}$$

In an external magnetic field, the average energy $\tilde{E}$ is the sum of $\langle E_{\mathrm{i}}\rangle$ and the average energy of the free spin system, $-MB$. In this approximation, the energy of a microscopic state depends only on M, and so the entropy, expressed as a function of M, is the same as before:

$$S\left(M\right) = k_{\mathrm{B}}N\left[\ln 2 - \frac{1}{2}\left(1+x\right)\ln\left(1+x\right) - \frac{1}{2}\left(1-x\right)\ln\left(1-x\right)\right] . \tag{7.30}$$

Thus the free energy $F(B,T,M)$ is given as follows as a function of B, T, M:

$$\begin{aligned} F(B,T,M) &= \langle E_{\mathrm{i}}\rangle - MB - TS \\ &= -\frac{z}{2}JNx^2 - MB \\ &\quad -Nk_{\mathrm{B}}T\left[\ln 2 - \frac{1}{2}\left(1+x\right)\ln\left(1+x\right) - \frac{1}{2}\left(1-x\right)\ln\left(1-x\right)\right] . \end{aligned} \tag{7.31}$$

Of the three variables B, T, and M, the first two, B and T, are external variables or constraints that we can fix arbitrarily. On the other hand, we cannot fix M; it is determined by the system. In accordance with the general principle that the most probable state is realized, the value of M at which the free energy becomes a minimum is also realized. Therefore $M(B,T)$ is determined by the condition that $\partial F/\partial M = 0$. Namely,

$$0 = \left(\frac{\partial F}{\partial M}\right)_{B,T} = -zJ\frac{x}{\mu} - B + \frac{k_{\mathrm{B}}T}{2\mu}\ln\left(\frac{1+x}{1-x}\right)$$
$$= -B_{\mathrm{eff}} + \frac{k_{\mathrm{B}}T}{2\mu}\ln\left(\frac{1+x}{1-x}\right) . \tag{7.32}$$

Here we have defined the effective magnetic field as $B_{\mathrm{eff}} \equiv B + zJx/\mu$, so that this equation has the same form as (7.8). Thus, the magnetization is

$$x = \frac{M}{M_{\mathrm{max}}} = \tanh\left(\frac{\mu B_{\mathrm{eff}}}{k_{\mathrm{B}}T}\right) . \tag{7.33}$$

This equation for $x = M/M_{\mathrm{max}}$ has the unknown x on the right-hand side also. Therefore, we must determine x self-consistently. In the mean-field approximation, we usually obtain this kind of equation, called a *self-consistent equation.*

Before solving this equation, we analyze the meaning of B_{eff}. A parallel link has an energy $-J$. We can imagine that the nearest link acts as if it induces an internal magnetic field $B_{\mathrm{int}} = J/\mu$ at the position of the neighboring spin. The interaction energy between the spin and B_{int} then reproduces the interaction energy $-J$. When there are z neighboring spins, there are zN_+/N up spins and zN_-/N down spins among these z spins on average. The total internal magnetic field is then

$$B_{\mathrm{int}} = z\frac{J}{\mu}\left(\frac{N_+}{N} - \frac{N_-}{N}\right) = z\frac{J}{\mu}x . \tag{7.34}$$

Adding the external field, we obtain the effective field given in (7.33), i.e. $B + B_{\mathrm{int}} = B_{\mathrm{eff}}$.

7.3.3 Solution of the Self-Consistent Equation

First we consider the situation in which there is no external magnetic field. In this case the self-consistent equation becomes

$$\frac{M}{M_{\mathrm{max}}} = \tanh\left(\frac{zJ}{k_{\mathrm{B}}T}\frac{M}{M_{\mathrm{max}}}\right) . \tag{7.35}$$

This equation always has a solution $M = 0$. This is the paramagnetic state. However, we have additional solutions at low temperature. In Fig. 7.6a, we have plotted the right-hand side of (7.35) as a function of M/M_{max}. The solution is given by the intersection of this curve with the dashed line, which represents the left-hand side, M/M_{max}. We define a critical temperature $T_{\mathrm{c}} \equiv$

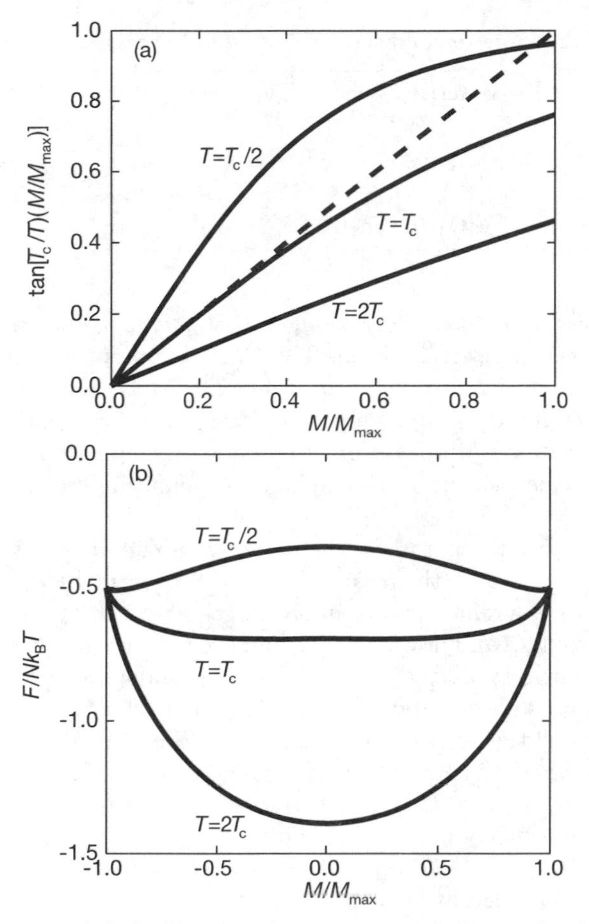

Fig. 7.6. Behavior of the self-consistent equation and F at several temperatures. In (**a**), the right-hand side of (7.35) is plotted as a function of $M/M_{\max}$. In (**b**), $F(0, T, M)$ is plotted as a function of $M/M_{\max}$. In these graphs, $T_c \equiv zJ/k_B$

zJ/k_B. When $T > T_c$, there is only one crossing, at the origin. At $T = T_c$, the solid curve touches the dashed line at the origin; at $T < T_c$, the slope of the solid curve near the origin exceeds unity, and an additional crossing appears at nonzero $M/M_{\max}$.

The self-consistent equation represents the condition that the free energy has an extreme value. Therefore, the solution can be found also from the plot of the free energy shown in Fig. 7.6b. At $T > T_c$, there is only one minimum, at $M/M_{\max} = 0$. At $T = T_c$, this minimum becomes very flat: around the minimum, $F \propto (M/M_{\max})^4$. Finally, at $T < T_c$, the extremum at $M/M_{\max} = 0$ changes into a maximum, and the minimum at $T > T_c$ separates into two minima situated symmetrically around the origin at nonzero $|M/M_{\max}|$. Therefore, when the temperature is lower than the critical temperature, a state with a nonzero magnetization is realized even without a magnetic field.

Table 7.2. Curie temperatures T_c for typical ferromagnetic materials

Material	Curie temperature
Fe	1043 K
Co	1400 K
Ni	637 K
Fe_3O_4 (magnetite)	860 K

The critical temperature is also called the *Curie temperature* in the case of the ferromagnetic phase transition. The Curie temperatures of some typical ferromagnetic materials are listed in Table 7.2. These temperatures suggest values of J several orders of magnitude larger than the values that can be estimated from the dipole–dipole interaction between magnetic moments. This is experimental evidence that J originates from a quantum mechanical exchange interaction.

Below the Curie temperature, there are two possibilities, either $M > 0$ or $M < 0$, but only one of them is realized.[7] At $T > T_c$ the two spin states, up and down, are symmetric and interchangeable. Each atomic moment can fluctuate between up and down equally, and the total magnetization fluctuates around the origin. At $T < T_c$ this symmetry is spontaneously broken. Once either of these possible orientations has been chosen, thermal fluctuations of M occur only around the corresponding minimum. We call this lack of symmetry below T_c *spontaneous symmetry breaking*, and call the change in the state at the critical temperature T_c a *phase transition*. The ferromagnetic state at $T < T_c$ is created by cooperation of spins. At higher temperatures this order is destroyed, since the $-ST$ term in the free energy favors a higher-entropy state, which is obtained at smaller M.

The temperature dependence of M obtained from (7.35) is shown in Fig. 7.7. Around $T \sim T_c$, the magnetization is small. Therefore, the self-consistent equation can be expanded around $x \equiv M/M_{\text{max}}$:

$$\frac{M}{M_{\text{max}}} \equiv x = \tanh\left(\frac{T_c}{T}x\right) \simeq \frac{T_c}{T}x - \frac{1}{3}\left(\frac{T_c}{T}x\right)^3 , \tag{7.36}$$

i.e.

$$\frac{1}{3}\left(\frac{T_c}{T}x\right)^3 = \left(\frac{T_c}{T} - 1\right)x . \tag{7.37}$$

[7] Which of these two states is realized depends on subtle details of the history of the system. For example, even at $T > T_c$, M has tiny fluctuations. The sign of M when the temperature of the system passes through the critical temperature will determine the sign of the magnetization at $T < T_c$.

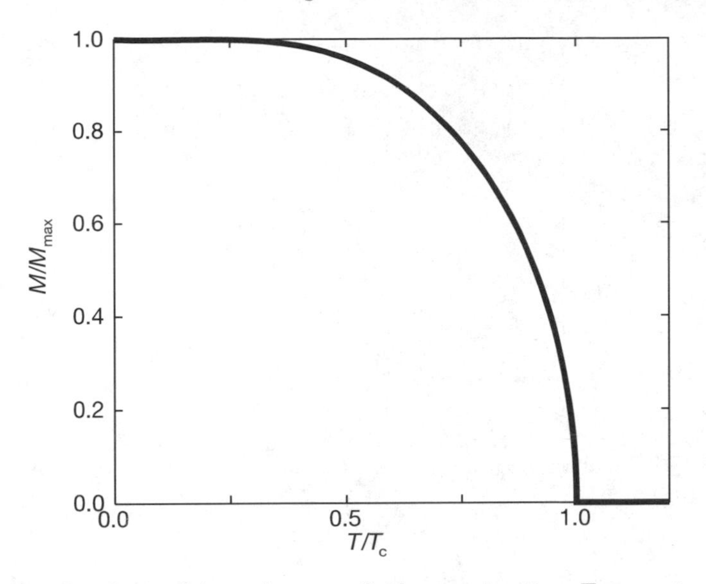

Fig. 7.7. Results obtained from the mean-field approximation. Temperature dependence of magnetization

From this equation, the magnetization just below the Curie temperature is given by[8]

$$\frac{M}{M_{\max}} \simeq \sqrt{3\frac{T_c - T}{T_c}} . \tag{7.38}$$

On the other hand, $M/M_{\max}$ when $T \to 0$ has the following temperature dependence:

$$\begin{aligned}\frac{M}{M_{\max}} \equiv x &= \tanh\left(\frac{T_c}{T}x\right) \\ &\simeq 1 - 2\exp\left(-2\frac{T_c}{T}x\right) \\ &\simeq 1 - 2\exp\left(-2\frac{T_c}{T}\right) .\end{aligned} \tag{7.39}$$

7.3.4 Entropy and Heat Capacity

The entropy predicted by the mean-field theory is obtained by putting the self-consistent solution for $x = M/M_{\max}$ into (7.30). The result at $B = 0$ is

[8] Around the Curie temperature, $x^2 = (M/M_{\max})^2$ can be expanded as a power series in the reduced temperature $t \equiv (T - T_c)/T_c$. Equation (7.38) shows the leading-order term in t. In order to obtain the next-order term, we need to expand $\tanh(T_c x/T)$ up to the fifth-order term in $(T_c x/T)$. Then, up to the order t^2, x^2 is given by $x^2 \simeq -3t - (12/5)t^2$.

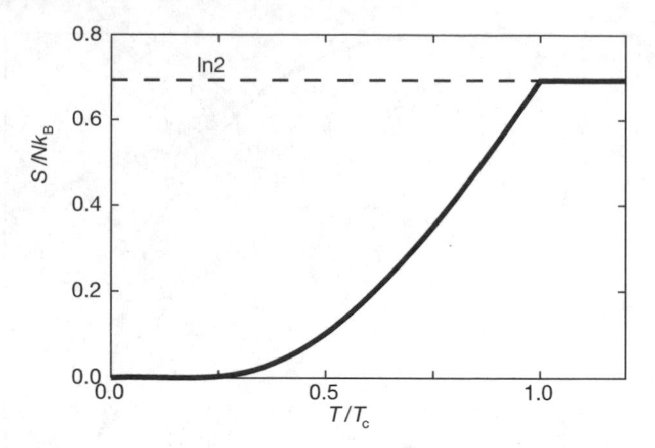

Fig. 7.8. Results obtained from the mean-field approximation. Temperature dependence of the entropy

shown in Fig. 7.8. It increases gradually with temperature, and saturates at the transition temperature.

Next we calculate the heat capacity at $B = 0$. This can be calculated either from the entropy using $C = T\,\partial S/\partial T$ or from the internal energy using $C = \partial U/\partial T$. The internal energy at $B = 0$ is

$$U = -\frac{z}{2}JN\left(\frac{M}{M_{\max}}\right)^2 . \tag{7.40}$$

At $T > T_c$ and $B = 0$, M is equal to 0. Hence $U = 0$ and $C = 0$. This result can also be obtained from the entropy. At $T < T_c$ and $B = 0$, a simple expression for the the heat capacity is obtained from the internal energy:

$$C = -zJN\frac{M}{M_{\max}^2}\frac{\partial M}{\partial T} . \tag{7.41}$$

The result is shown in Fig. 7.9. A numerical solution of the self-consistent equation has been used to obtain the temperature dependence of M. The value at $T = T_c$ can be calculated analytically from (7.38). Just below the Curie temperature, the energy is

$$U = \frac{3}{2}k_BN(T - T_c) . \tag{7.42}$$

Thus, $C(T_c) = (3/2)k_BN$.[9]

[9] If we use the expression for x^2 given in footnote 8, $C(T) = (3/2)k_BN[1+(8/5)(T-T_c)/T_c]$ is obtained. This expression reproduces the linear decrease of C for $T < T_c$ in Fig. 7.9.

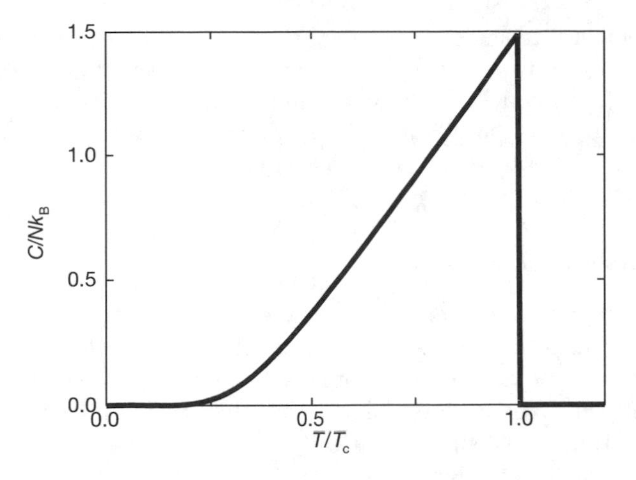

Fig. 7.9. Results obtained from the mean-field approximation. Temperature dependence of the heat capacity

7.3.5 Susceptibility

The susceptibility $\chi(T)$ at $T > T_c$ is given by

$$\begin{aligned}\chi(T) &= \mu_0 \left(\frac{\partial M}{\partial B}\right)_T = \mu_0 M_{\max} \frac{\partial}{\partial B} \tanh\left(\frac{\mu B_{\text{eff}}}{k_B T}\right) \\ &= \mu_0 M_{\max} \frac{\mu}{k_B T} \operatorname{sech}^2\left(\frac{\mu B_{\text{eff}}}{k_B T}\right) \frac{\partial B_{\text{eff}}}{\partial B} \\ &= \frac{\mu \mu_0 M_{\max}}{k_B T} \operatorname{sech}^2\left(\frac{\mu B_{\text{eff}}}{k_B T}\right) \left[1 + z \frac{J}{\mu} \frac{1}{M_{\max}} \left(\frac{\partial M}{\partial B}\right)_T\right] \\ &= \frac{\mu \mu_0 M_{\max}}{k_B T} \operatorname{sech}^2\left(\frac{\mu B_{\text{eff}}}{k_B T}\right) + \frac{zJ}{k_B T} \chi \operatorname{sech}^2\left(\frac{\mu B_{\text{eff}}}{k_B T}\right) . \end{aligned} \tag{7.43}$$

From this equation, we obtain

$$\chi = \frac{\dfrac{\mu \mu_0 M_{\max}}{k_B T} \operatorname{sech}^2\left(\dfrac{\mu B_{\text{eff}}}{k_B T}\right)}{1 - \dfrac{zJ}{k_B T} \operatorname{sech}^2\left(\dfrac{\mu B_{\text{eff}}}{k_B T}\right)} . \tag{7.44}$$

In the paramagnetic phase, $\operatorname{sech}^2(\mu B_{\text{eff}}/k_B T) \to 1$ as $B \to 0$. Therefore, χ can be expressed as

$$\chi(T) = \frac{\mu \mu_0 M_{\max}}{k_B (T - T_c)} = \mu_0 \frac{\mu^2 N}{k_B} \frac{1}{T - T_c} . \tag{7.45}$$

This temperature dependence is called the *Curie–Weiss law*. The divergence of χ at $T = T_c$ is a general property of the ferromagnetic phase transition, not a special consequence of the mean-field approximation. It means that a nonzero magnetization is induced by an infinitesimally weak magnetic field at the Curie temperature. At $T < T_c$, no magnetic field is needed to obtain a nonzero M.

7.3.6 Domain Structure

Iron or steel at room temperature is in the ferromagnetic phase, but a block of iron as a whole does not usually have a nonzero magnetization. This is because it consists of many small ferromagnetic domains, in which the magnetic moments of the domains are aligned such that they cancel when the total moment is taken, as shown schematically in Fig. 7.10. We have remarked that the dipole–dipole interaction between spins is too small to be the origin of the ferromagnetic phase. However, once a system is in the ferromagnetic phase, this interaction becomes important. This is because the dipole interaction is long-ranged, and aligned spins make a large total moment. When two bar magnets are put together, their north poles are attracted to the south poles of the other magnet. As a result, they will stick together so that the north pole of each magnet is in contact with the south pole of the other magnet. As a result, the total moment is reduced. In this configuration, the energy of the magnetic field is reduced, and the two magnets are in their most stable state. Similarly, a block of iron is divided into many domains and the moments of the domains cancel to reduce the energy of the magnetic field. If the magnetic moments of the domains are aligned, the block becomes a permanent magnet.

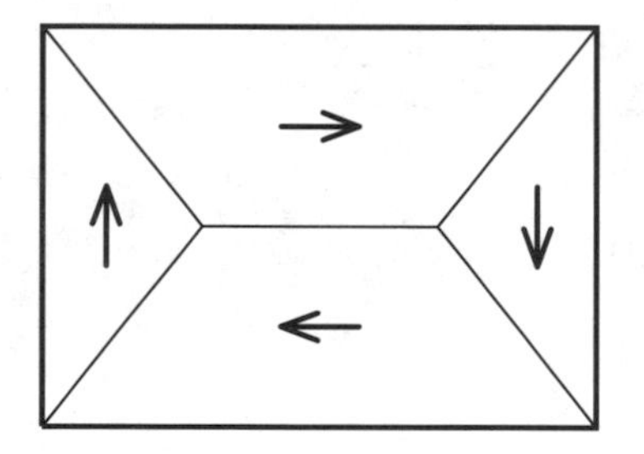

Fig. 7.10. An example of a magnetic-domain structure. In this case the system is divided into four domains. The magnetic moment in each domain is shown by an *arrow*

7.4 The One-Dimensional Ising Model

7.4.1 Free Energy

In the previous section, we investigated the Ising model by use of the mean-field theory. The free energy obtained there was an approximation. An exact calculation of the free energy of an interacting system is impossible in general. However, in one dimension, the free energy of certain systems can be calculated exactly. The Ising model is one such system, and so here we calculate its free energy exactly.

We investigate a system of unit length consisting of N spins aligned linearly. The ith spin is denoted by σ_i, which is equal to either $+1$ (up) or -1 (down). We impose a periodic boundary condition so that $\sigma_{N+1} = \sigma_1$. The spins interact with their nearest-neighbor spins. A microscopic state given by a spin configuration $(\sigma_1, \sigma_2, \cdots, \sigma_N)$ has the following energy E:

$$E(\{\sigma_i\}) = -J \sum_{i=1}^{N} \sigma_i \sigma_{i+1} - \mu B \sum_{i=1}^{N} \sigma_i, \qquad \sigma_i = \pm 1 , \tag{7.46}$$

where the magnitude of the magnetic moment μ has been included in the coupling constant J, and B is the external magnetic field. The partition function is the following sum over the microscopic states:

$$\begin{aligned} Z &= \sum_{\{\sigma_i\}} \mathrm{e}^{-\beta E(\{\sigma_i\})} \\ &= \sum_{\sigma_1} \sum_{\sigma_2} \cdots \sum_{\sigma_N} \prod_{i=1}^{N} A(\sigma_i, \sigma_{i+1}) , \end{aligned} \tag{7.47}$$

where

$$A(\sigma_i, \sigma_{i+1}) = \exp\left[\beta J \sigma_i \sigma_{i+1} + \frac{1}{2}\beta\mu B \sigma_i + \frac{1}{2}\beta\mu B \sigma_{i+1}\right] . \tag{7.48}$$

Let us consider the sum over the ith spin σ_i:

$$\sum_{\sigma_i} A(\sigma_{i-1}, \sigma_i) A(\sigma_i, \sigma_{i+1}) . \tag{7.49}$$

We notice that this can be considered as a matrix product of a two-by-two matrix A with itself, where

$$\mathsf{A} = \begin{pmatrix} A(+,+) & A(+,-) \\ A(-,+) & A(-,-) \end{pmatrix} . \tag{7.50}$$

To simplify the treatment, we introduce the notation $K = \beta J$ and $b = \beta \mu B$. The matrix elements of A are

$$A(+,+) = \exp\left(\beta J + \frac{1}{2}\beta\mu B + \frac{1}{2}\beta\mu B\right) = \mathrm{e}^{K+b} , \tag{7.51}$$

$$A(+,-) = A(-,+) = \exp(-\beta J) = \mathrm{e}^{-K} , \tag{7.52}$$

$$A(-,-) = \exp\left(\beta J - \frac{1}{2}\beta\mu B - \frac{1}{2}\beta\mu B\right) = \mathrm{e}^{K-b} . \tag{7.53}$$

Thus the two-by-two matrix A has the following form:

$$\mathsf{A} = \begin{pmatrix} \mathrm{e}^{K+b} & \mathrm{e}^{-K} \\ \mathrm{e}^{-K} & \mathrm{e}^{K-b} \end{pmatrix} . \tag{7.54}$$

This matrix A is called a *transfer matrix.*

The partition function can now be rewritten as follows:

$$Z = \mathrm{Tr}\left(\mathsf{A}^N\right) . \tag{7.55}$$

Since the matrix A is real and symmetric, it can be diagonalized:

$$\mathsf{U}^{-1}\mathsf{A}\mathsf{U} = \begin{pmatrix} \lambda_+ & 0 \\ 0 & \lambda_- \end{pmatrix} . \tag{7.56}$$

The partition function can be calculated as follows:

$$\mathrm{Tr}\left(\mathsf{A}^N\right) = \mathrm{Tr}\left(\mathsf{U}^{-1}\mathsf{A}\mathsf{U}\mathsf{U}^{-1}\mathsf{A}\mathsf{U}\cdots\mathsf{A}\mathsf{U}\right) = \lambda_+^N + \lambda_-^N . \tag{7.57}$$

Since N is a macroscopic number, if $\lambda_+ > \lambda_-$ then $\lambda_+^N \gg \lambda_-^N$, and $Z = \lambda_+^N$ for $N \to \infty$. The eigenvalues $\lambda_\pm$ can be calculated as the solutions of the following equation:

$$\begin{vmatrix} \lambda - \mathrm{e}^{K+b} & -\mathrm{e}^{-K} \\ -\mathrm{e}^{-K} & \lambda - \mathrm{e}^{K-b} \end{vmatrix} = \lambda^2 - 2\lambda\mathrm{e}^K \cosh b + 2\sinh 2K = 0 . \tag{7.58}$$

The larger solution λ_+ is

$$\lambda_+ = \mathrm{e}^K \cosh b + \sqrt{\mathrm{e}^{2K}\cosh^2 b - 2\sinh 2K} . \tag{7.59}$$

When $B = 0$, i.e. when $b = 0$,

$$\lambda_+ = \mathrm{e}^K + \mathrm{e}^{-K} = 2\cosh K , \tag{7.60}$$

and the free energy is

$$F = -Nk_\mathrm{B}T \ln\left(2\cosh\beta J\right) . \tag{7.61}$$

7.4.2 Entropy and Heat Capacity

The entropy at $B = 0$ can be calculated by differentiating the free energy with respect to the temperature:

$$S = -\frac{\partial F}{\partial T} = Nk_\mathrm{B}\left[\ln\left(2\cosh\frac{J}{k_\mathrm{B}T}\right) - \frac{J}{k_\mathrm{B}T}\left(\tanh\frac{J}{k_\mathrm{B}T}\right)\right] . \tag{7.62}$$

The result is shown in Fig. 7.11.

The internal energy and the heat capacity are

$$E = F + ST = -NJ \tanh \frac{J}{k_{\mathrm{B}}T} \tag{7.63}$$

and

$$C = \frac{\mathrm{d}E}{\mathrm{d}T} = \frac{NJ^2}{k_{\mathrm{B}}T^2} \operatorname{sech}^2 \frac{J}{k_{\mathrm{B}}T} \,. \tag{7.64}$$

The heat capacity can also be calculated from $C = T\,\partial S/\partial T$. The temperature dependence is shown in Fig. 7.12. The heat capacity has the same temperature

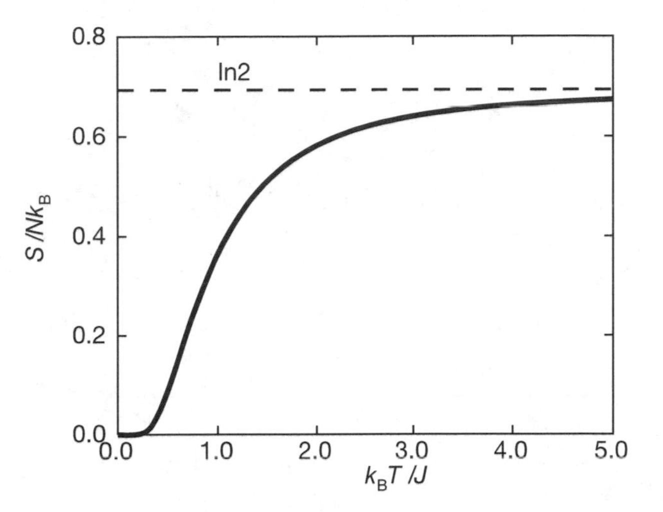

Fig. 7.11. Exact result for one-dimensional Ising model: entropy plotted as a function of $k_{\mathrm{B}}T/J$

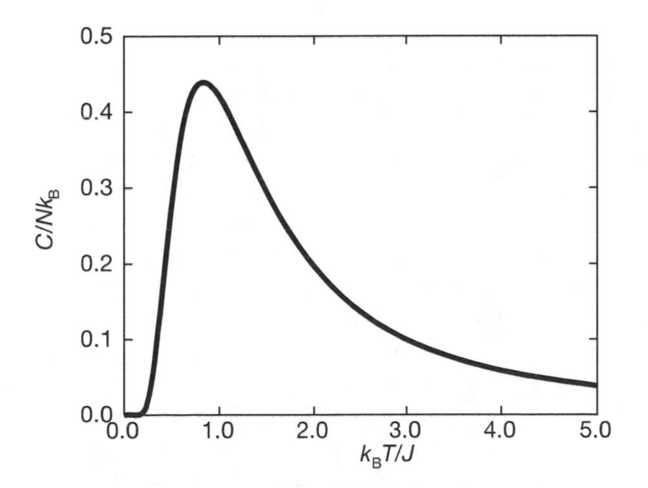

Fig. 7.12. Exact result for one-dimensional Ising model: heat capacity plotted as a function of $k_{\mathrm{B}}T/J$

dependence as that of a system of free spins in a magnetic field $B = J/\mu$. Compare this result with that obtained from the mean-field approximation, which has a discontinuity at $T_\mathrm{c} = 2J/k_\mathrm{B}$ in the present notation.

7.4.3 Magnetization and Susceptibility

Here we allow the magnetic field to have a nonzero value; that is, $b = \beta\mu B \neq 0$. The free energy F and the magnetization M are now given by

$$F = -Nk_\mathrm{B}T \ln\left[\mathrm{e}^K \cosh b + \sqrt{\mathrm{e}^{2K}\cosh^2 b - 2\sinh 2K}\right] \tag{7.65}$$

and

$$\begin{aligned} M &= -\frac{\partial F}{\partial B} = -\frac{\partial F}{\partial b}\beta\mu \\ &= \beta\mu N k_\mathrm{B}T \frac{\mathrm{e}^K \sinh b + \dfrac{\mathrm{e}^{2K}\sinh b \cosh b}{\sqrt{\mathrm{e}^{2K}\cosh^2 b - 2\sinh 2K}}}{\mathrm{e}^K \cosh b + \sqrt{\mathrm{e}^{2K}\cosh^2 b - 2\sinh 2K}} \,. \end{aligned} \tag{7.66}$$

In the limit of a weak magnetic field, $\sinh b \simeq b$, $\cosh b \simeq 1$, and

$$M \simeq N\mu \frac{\mathrm{e}^K + \mathrm{e}^{2K}/\mathrm{e}^{-k}}{\mathrm{e}^K + \mathrm{e}^{-K}} \times b = \frac{N\mu^2}{k_\mathrm{B}T} \times \mathrm{e}^{2K} B = \frac{N\mu^2}{k_\mathrm{B}T}\mathrm{e}^{2J/k_\mathrm{B}T} B \,. \tag{7.67}$$

Thus there is no spontaneous magnetization: when $B = 0$, $M = 0$. The susceptibility is given by

$$\chi = \mu_0 \frac{N\mu^2}{k_\mathrm{B}T}\mathrm{e}^{2J/k_\mathrm{B}T} \,. \tag{7.68}$$

The susceptibility of a free spin system, $\mu_0 N\mu^2/k_\mathrm{B}T$, is enhanced by a factor $\mathrm{e}^{2J/k_\mathrm{B}T}$. The temperature dependence is shown in Fig. 7.13.

The result of this exact calculation has shown that there is no phase transition in this one-dimensional Ising model. This is a special feature of a one-dimensional system. The order in a one-dimensional system is quite fragile with respect to thermal fluctuations. Since each spin has only two nearest neighbors, a flip of a single spin can cause a spin flip of a neighbor with a probability of 1/2, and so a spin flip can propagate and may cause flips of the majority of spins. On the other hand, propagation of a spin flip is unlikely in two dimensions if the spins are already mostly aligned. In fact, it has been shown that there is a phase transition in the two-dimensional Ising model, as we shall see in Chap. 9. The transition temperature in this case is given by $k_\mathrm{B}T_\mathrm{c} = J/\sinh^{-1} 1 = 1.13J$ for a square lattice. The investigation of a one-dimensional system is sometimes fruitful, because we can obtain exact solutions, but we must be aware that the results may be qualitatively different from those for real three-dimensional systems.

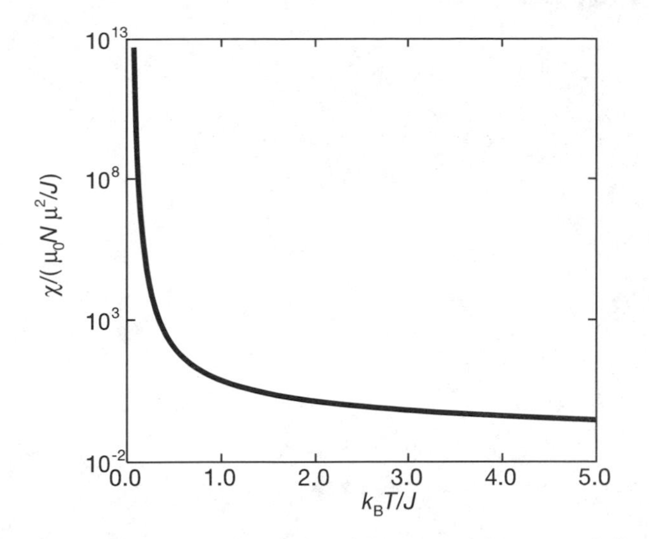

Fig. 7.13. Exact result for one-dimensional Ising model: susceptibility plotted as a function of $k_\mathrm{B}T/J$. Note that the *vertical axis* is logarithmic

Exercise 15. In order to estimate the strength of the magnetic dipole interaction between atoms, consider the interaction of the spin magnetic moments of two electrons separated by a typical interatomic distance of 0.5 nm. Calculate the interaction energy for two cases: (1) the spins are parallel to the line joining the electrons and point in the same direction, and (2) the spins are parallel to the line joining the electrons but point in opposite directions. Express the interaction energy as a temperature.

Part III

More Advanced Topics

8

First-Order Phase Transitions

In the previous chapter we have seen that a magnetic material undergoes a phase transition between a paramagnetic phase and a ferromagnetic phase as the temperature is varied. There are various other kinds of phase transitions in nature apart from this ferromagnetic phase transition. They can be categorized into two kinds. One kind, *first-order transitions*, is characterized by a discontinuous change in the internal energy. The general behavior of this kind of transition is discussed in this chapter. The other kind is *second-order transitions*, in which the internal energy changes continuously across the transition point. This kind of transition is discussed in Chap. 9.

8.1 The Various Phases of Matter

Water has at least three different phases: liquid, solid (ice), and gas (vapor).[1] At any given temperature and pressure, one of these phases exists, and for some special combinations of temperature and pressure, two or all three phases coexist. Likewise, other substances exist in various phases. For example, nitrogen at atmospheric pressure is in the gas phase at room temperature, but it liquefies at 77.35 K and solidifies at 63.29 K. Helium gas is hard to liquefy, but the helium isotope of mass number 4 liquefies at 4.2 K at atmospheric pressure. Liquid helium can be divided further into a normal fluid phase at $T > T_\lambda$ and a superfluid phase at $T < T_\lambda$, where T_λ, the lambda transition temperature, depends on the pressure, and is 2.17 K under the saturated vapor pressure. A solid can also have different phases. For example, as explained in Chap. 7, iron is in a paramagnetic phase at $T > T_c$ and a ferromagnetic phase at $T < T_c$.

As the pressure or temperature changes, one phase may be replaced by another phase. This change of phase is called a *phase transition*. Phase transitions can be divided into two classes. In a first-order phase transition, the internal energy per mole changes by a nonzero amount. On the

[1] Strictly speaking, ice has several different phases.

other hand, in a second-order phase transition, the internal energy changes continuously through the transition point. The phase transitions between a gas and a liquid, between a liquid and a solid, and between a solid and a gas are usually of first order. The magnetic transition and the normal-to-superfluid transition of liquid helium are of second order. In this chapter, we consider first-order phase transitions; second-order transitions are treated in Chap. 9.

The reason why a gas liquefies or solidifies is the attractive interaction between its atoms or molecules. Without such an interaction, a gas would behave as an ideal gas, and would remain a gas at any temperature above absolute zero. The interaction between the atoms or molecules is usually caused by their electric dipole moments. A water molecule has a permanent electric dipole moment, since the three atoms in the H–O–H molecule are not in a straight line, and the hydrogen atoms are slightly positively charged and the oxygen atom is slightly negatively charged. Therefore, between suitably oriented molecules, there is an attractive interaction. The nitrogen molecule, oxygen molecule, and helium atom do not have a permanent electric dipole moment. However, electron motion in a molecule causes a temporally varying dipole moment, and a fluctuating electric field around the molecule. This fluctuating electric field can polarize nearby molecules and induce a dipole moment in them, as shown in Fig. 8.1. Once a moment has been induced, it in turn supports the dipole moment in the original molecule, and so the average dipole moment of this pair of molecules becomes nonzero, and an attractive interaction acts between the molecules. The potential energy between these induced moments varies as r^{-6}, and the force between them varies as r^{-7}.

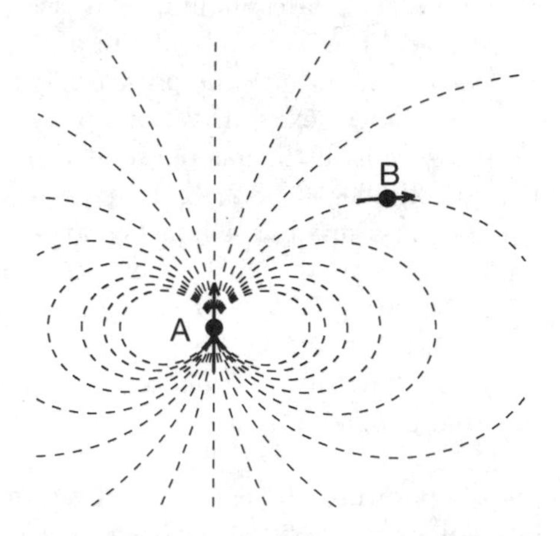

Fig. 8.1. Interaction between neutral molecules or atoms. Even if molecule A is electrically neutral and does not have a permanent electric dipole, it has a fluctuating electric dipole moment, which polarizes a nearby molecule B and induces an electric dipole in it. These electric dipoles causes an attractive interaction between them

The attractive interaction is replaced by a strong repulsive interaction when the molecules begins to overlap with each other. This repulsive part is called the hard-core repulsion between the molecules. The total interaction, which is repulsive at short distances and attractive at long distances, is called the *van der Waals interaction.* The typical behavior of the van der Waals interaction potential is shown in Fig. 8.2.

At zero temperature, the lowest-energy state is realized. If there are only two molecules, they will form a bound state where the distance between the molecules is given by the position of the minimum of the interaction potential. Therefore, they can no longer behave as free molecules. Adding another molecule will result in a bound state of three molecules. If we imagine the addition of more molecules, we can arrive at a macroscopic system. Here the state of lowest potential energy will be a crystal of molecules, in which the molecules are placed in a lattice of some kind.

At zero temperature, there are small vibrations of the molecules due to the quantum mechanical zero-point motion. For most substances, this zero-point motion is not important. However, the case of helium atoms is an exception. In this substance, the zero-point motion is so large that the crystal melts, and helium remains liquid unless a pressure of about 25 atm is applied. For other solids, the lattice vibrations, which give rise to the lattice specific heat considered in Chap. 5, become more violent as the temperature is raised. At some particular temperature the crystal melts into a liquid state. The liquid state has a similar density to the solid state, and so the distance between the molecules is nearly the same as in the solid; therefore, the liquid is also bound by the van der Waals interaction. Melting occurs when the temperature is much higher than the Debye temperature, and so the equipartition law is

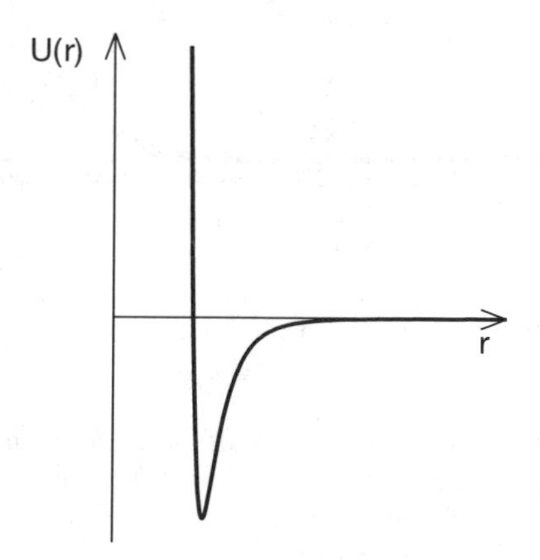

Fig. 8.2. The van der Waals potential. This has a strongly repulsive part at short distances (hard-core part), and an attractive part due to dipole–dipole interaction which decreases as r^{-6} at long distances

expected to be satisfied. This means that each molecule has an average kinetic energy of

$$\left\langle \frac{1}{2} m v^2 \right\rangle \simeq \frac{3}{2} k_{\mathrm{B}} T \,. \tag{8.1}$$

When this average energy becomes of the same order as the binding energy due to the van der Waals interaction, the molecules are no longer bound into a liquid, and a gas phase is obtained.

This is a rough scenario for the phase transitions of a substance that is in the gas phase at room temperature.[2] The strength of the attraction part depends on the polarizability of the molecules, which is a measure of how easily a molecule is polarized. The electrons in a helium atom form a closed shell, and it is hard to change the configuration of the electrons and therefore it is hard to polarize a helium atom. Because of this, the interaction between helium atoms is weak, and the liquefaction temperature is low. Compared with helium, oxygen molecules and nitrogen molecules have a larger polarizability, and so oxygen and nitrogen liquefy at higher temperatures. Needless to say, water, which has a permanent electric dipole moment, has higher transition temperatures than those of oxygen and nitrogen.

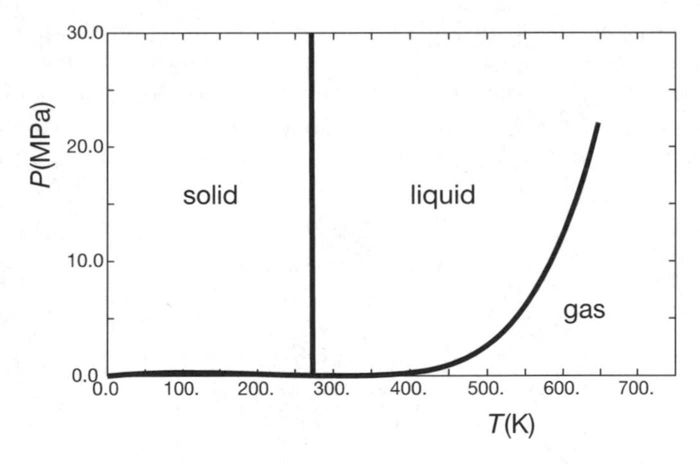

Fig. 8.3. Phase diagram of water. There are two special points: the triple point at $T = 273.16\,\mathrm{K}$ and $P = 611.66\,\mathrm{Pa}$, and the critical point at $T = 647.31\,\mathrm{K}$ and $P = 22.106\,\mathrm{MPa}$

[2] There are also substances which are in a solid phase at room temperature. In these substances, the attractive interaction between the atoms is stronger than the van der Waals attraction. In ionic crystals, such as NaCl, positively charged Na^+ ions and negatively charged Cl^- ions attract as a result of the Coulomb interaction. In metals, the positive ions are bound together by the negative conduction electrons (metallic bonding). The carbon atoms in diamond are bound together by strong covalent bonds.

The temperature of a phase transition point is affected by the pressure. Therefore, a *phase diagram* can be drawn in the plane of temperature and pressure. As an example of a phase diagram, that of water is shown in Fig. 8.3. This tells us which phase exists at any given temperature and pressure. There are two special points in the phase diagram. One is the *triple point*, at which the solid, liquid, and gas phases can coexist. This occurs at $T = 273.16\,\mathrm{K}$ and $P = 611.66\,\mathrm{Pa}$.[3] Another is the *critical point*, beyond which the line separating the gas and liquid phases disappears. This is the point at $T = 647.31\,\mathrm{K}$ and $P = 22.106\,\mathrm{MPa}$. In the following sections, we discuss first-order phase transitions with the help of statistical physics.

8.2 System in a Heat Bath at Fixed *P* and *T*

For the discussion of first-order phase transitions, we need to consider a system of N particles in a heat bath, at a constant temperature T and pressure P. Under these conditions, every possible microscopic state of the system, with energy E and volume V, is realized with a probability determined by the law of statistical physics, namely the principle of equal probability. We can repeat the discussion in Chap. 3 with slight modifications to obtain this probability.

Let the total energy of the system (system I) plus the heat bath (system II) be E_t, and let the total volume be V_t. The probability that system I is in a state with energy E_I and volume V_I is proportional to the total number of microscopic states of the whole system that allow system I to be in that state. That is, the probability is proportional to the number of microscopic states in which the heat bath has an energy $E_\mathrm{t} - E_\mathrm{I}$ and volume $V_\mathrm{t} - V_\mathrm{I}$. This number is given by the entropy of the heat bath $S_\mathrm{II}(E, V)$. We obtain

$$
\begin{aligned}
\text{Probability} &\propto \frac{\exp\left[S_\mathrm{II}(E_\mathrm{t} - E_\mathrm{I}, V_\mathrm{t} - V_\mathrm{I})/k_\mathrm{B}\right]}{\exp\left[S_\mathrm{II}(E_\mathrm{t}, V_\mathrm{t})/k_\mathrm{B}\right]} \\
&= \exp\left\{\frac{1}{k_\mathrm{B}}\left[S_\mathrm{II}(E_\mathrm{t} - E_\mathrm{I}, V_\mathrm{t} - V_\mathrm{I}) - S_\mathrm{II}(E_\mathrm{t}, V_\mathrm{t})\right]\right\} \\
&\simeq \exp\left[\frac{1}{k_\mathrm{B}}\left(-E_\mathrm{I}\frac{\partial S_\mathrm{II}(E_\mathrm{t}, V_\mathrm{t})}{\partial E_\mathrm{t}} - V_\mathrm{I}\frac{\partial S_\mathrm{II}(Et, V_\mathrm{t})}{\partial V_\mathrm{t}}\right)\right] \\
&= \exp\left(-\frac{E_\mathrm{I}}{k_\mathrm{B}T} - \frac{PV_\mathrm{I}}{k_\mathrm{B}T}\right),
\end{aligned}
\tag{8.2}
$$

where T and P are the temperature and pressure of the heat bath.

[3] This pressure is so low that it looks as if the gas phase terminates at around 300 K in this figure. In fact, the gas phase still exists on the low-pressure side of the solid phase, and extends down to $T = 0\,\mathrm{K}$.

The relative probability that system I has an energy between E_{I} and $E_{\mathrm{I}} + \delta E_{\mathrm{I}}$ and a volume between V_{I} and $V_{\mathrm{I}} + \delta V_{\mathrm{I}}$ is then given by the density of states of system I, $\Omega_{\mathrm{I}}(E_{\mathrm{I}}, V_{\mathrm{I}}, N)$, or by the entropy of that system, $S_{\mathrm{I}}(E, V, N)$:

$$\begin{aligned} P_{\mathrm{rel}}(E_{\mathrm{I}}, V_{\mathrm{I}}, N)\, \mathrm{d}E_{\mathrm{I}}\, \mathrm{d}V_{\mathrm{I}} &\propto \Omega_{\mathrm{I}}(E_{\mathrm{I}}, V_{\mathrm{I}}, N)\, \mathrm{d}E_{\mathrm{I}}\, \mathrm{d}V_{\mathrm{I}} \exp\left(-\frac{E_{\mathrm{I}}}{k_{\mathrm{B}}T} - \frac{PV_{\mathrm{I}}}{k_{\mathrm{B}}T}\right) \\ &= \mathrm{d}E_{\mathrm{I}}\, \mathrm{d}V_{\mathrm{I}} \exp\left\{-\frac{1}{k_{\mathrm{B}}T}\left[E_{\mathrm{I}} + PV_{\mathrm{I}} - S_{\mathrm{I}}(E_{\mathrm{I}}, V_{\mathrm{I}}, N)T\right]\right\} . \end{aligned} \tag{8.3}$$

The partition function in this situation, $Y(T, P, N)$, is obtained by integrating this relative probability over the energy and volume:

$$Y(T, P, N) = \int_0^\infty \mathrm{d}V_{\mathrm{I}} \int_0^\infty \mathrm{d}E_{\mathrm{I}} \exp\left\{-\frac{1}{k_{\mathrm{B}}T}\left[E_{\mathrm{I}} + PV_{\mathrm{I}} - S_{\mathrm{I}}(E_{\mathrm{I}}, V_{\mathrm{I}}, N)T\right]\right\} . \tag{8.4}$$

Because we are considering a macroscopic system as in Chap. 3, the integrand, which is the probability of finding system I at E_{I} and V_{I}, should be sharply peaked at the minimum of $E_{\mathrm{I}} + PV_{\mathrm{I}} - S_{\mathrm{I}}T$, at $E^*(T, P, N)$ and $V^*(T, P, N)$. Therefore, the system will almost always be found with this energy and volume. The energy at the minimum E^* is the internal energy U of the system. Since $U + PV - ST = G(T, P, N)$ is the Gibbs free energy introduced in Sect. 3.5, what is shown here is that under this condition of given T and P, the state realized has the lowest Gibbs free energy with respect to E and V.

A relation between the partition function Y and the Gibbs free energy G can be obtained by evaluating Y through expanding the integrand around E^* and V^*. Since the integrand decreases rapidly around the peak, expansion up to the second order is sufficient. The integral then becomes Gaussian, and so the result is

$$\begin{aligned} Y(T, P, N) &= C(T, P, N) \exp\left\{-\frac{1}{k_{\mathrm{B}}T}\left[E^* + PV^* - S_{\mathrm{I}}(E^*, V^*, N)T\right]\right\} \\ &= C(T, P, N) \exp\left[-\frac{1}{k_{\mathrm{B}}T} G(T, P, N)\right] , \end{aligned} \tag{8.5}$$

where $C(T, P, N)$ is some function of T, P, and N, which is obtained from the Gaussian integral; an explicit expression is not needed here. We obtain

$$\begin{aligned} G(T, P, N) &= -k_{\mathrm{B}}T\left[\ln Y(T, P, N) - \ln C(T, P, N)\right] \\ &= -k_{\mathrm{B}}T \ln Y(T, P, N) . \end{aligned} \tag{8.6}$$

Since G and $\ln Y$ are quantities of order N, $\ln C = O(1)$ has been neglected in obtaining the second line of the equation above.

8.3 Coexistence of Phases

As shown in the previous section, at given T and P a system is in the phase which has the lowest value of $E + PV - S(E, V, N)T$, and this lowest value is equal to G. The phase may be a solid, liquid, or gas depending on the temperature and pressure. Experimentally, these phase transitions occur as the temperature or pressure is changed, and so the minimum should be transferred from one phase to another at the transition point. That is, the coexistence condition is the condition that the Gibbs free energy has the same value for the two phases. Since the Gibbs free energy divided by the number of molecules is the chemical potential $\mu(T, P) = G(T, P, N)/N$, we can also say that the chemical potentials of the two phases are equal when they coexist.

Let us see how the transfer of the minimum occurs. First, we consider the solid-to-liquid phase transition. In general, the liquid phase has a slightly larger molar volume and a slightly higher internal energy. The molecules in the liquid require a larger nearest-neighbor distance to move around, and so the volume expands, and the average potential energy of interaction is higher. The molecules or atoms in the liquid phase can be anywhere in the volume, but in the solid phase they can move only around their lattice points. Therefore, the number of possible microscopic states is much larger in the liquid phase than in the solid phase, and the entropy of the liquid phase should be larger. Near the phase transition point, there should be two local minima of $E+PV-S(E, V)T$ in the E–V plane. One of these minima corresponds to the solid phase and the other to the liquid phase. The quantity $E+PV$ is lower at the solid-phase local minimum, as explained above. Therefore, at lower temperatures, where the contribution ST is small, the solid phase has a lower value of $E+PV-ST$. As the temperature becomes higher, the liquid-phase minimum decreases faster than the solid-phase minimum owing to the larger value of S, and so a transfer of the minimum from one phase to the other occurs at some temperature.

Next we consider the liquid-to-gas phase transition. In this case the energy is higher in the gas phase because there is practically no energy gain from the attractive interaction of the molecules. The volume and entropy are much larger in the gas phase. Therefore, both temperature and pressure can be effective in causing a transition between the phases. The liquid phase is favored at high pressure because of the PV term. The gas phase is favored at high temperature because of the $-ST$ term. These tendencies are in accordance with our everyday experience.

The coexistence condition defines lines in the P–T plane. On the coexistence line, the volume and the internal energy of the system are somewhat arbitrary. For example, let us consider the case of a gas–liquid transition, where the system consists of one mole of molecules of some kind. When the system becomes a liquid it has a smaller molar volume V_{l}, and when it becomes a gas it has a larger molar volume V_{g}. At the coexistence point, the volume can take values between V_{l} and V_{g}. Namely, if a mass fraction x is in the liquid phase and the remaining fraction $1-x$ is in the gas phase, the total volume is

$V = xV_l + (1-x)V_g$, as shown in Fig. 8.4. What determines the fraction x is the total internal energy, given by $U = xU_l + (1-x)U_g$, which can be varied between U_l and U_g. Since the Gibbs free energies per mole are the same on the coexistence line, the differences in the molar entropy $\Delta S \equiv S_g - S_l$, in the molar volume $\Delta V \equiv V_g - V_l$, and in the molar internal energy $\Delta U \equiv U_g - U_l$ are related:

$$T\,\Delta S = \Delta U + P\,\Delta V\,. \tag{8.7}$$

When we add a quantity of heat Q to this system the total entropy increases by $T\,\delta S = Q$. Some of the liquid evaporates because of this heat, and the fraction x increases to $x + \delta x$, where $\delta x = \delta S/\Delta S$. The total energy then increases by $\delta U = \Delta U\,\delta x$, and the total volume increases by $\delta V = \Delta V\,\delta x$. The quantity of heat needed to convert one mole of liquid into one mole of gas, $Q_L = T\,\Delta S$, is called the *latent heat*. Since the entropy changes across the coexistence line, i.e. the line of the first-order phase transition, a latent heat is always needed for a first-order phase transition.

In equilibrium statistical physics, we consider equilibrium phases that are realized after waiting for a long time for the system to stabilize. Therefore, at given T and P there is only one stable phase, except on the coexistence line. However, in reality it can happen that a phase continues across the coexistence line. That is, a liquid may continue to be a liquid above the boiling point or below the freezing point. These phenomena are called *superheating* and *supercooling*, respectively, and can occur at any phase boundary. The reason for these phenomena is that there are two local minima of the quantity $E + PV - ST$ in the E–V plane. One of them, the true minimum, is realized in thermal equilibrium, but the other one can also be realized in a metastable state. In a supercooled or superheated state, the system is in

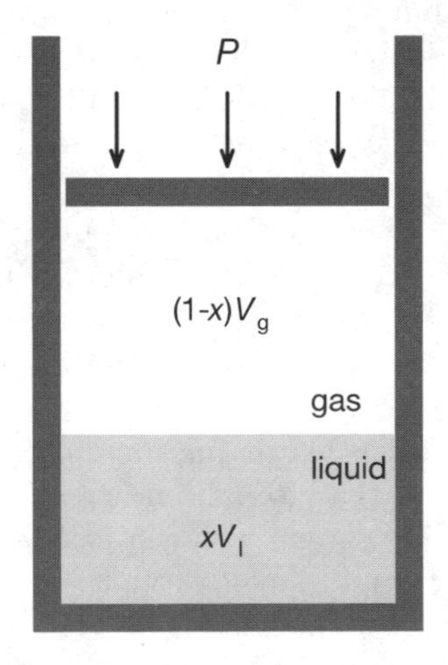

Fig. 8.4. Coexistence of liquid and gas phases. The system is assumed to be in a heat bath at temperature T and pressure P, on the coexistence line in the phase diagram. The total volume $xV_l + (1-x)V_g$ can be anywhere between V_l and V_g

such a metastable state. Interaction with the heat bath makes the system fluctuate around the corresponding minimum with a probability given by the canonical distribution. A rather large perturbation is needed from the heat bath to make the system move to a configuration around the other (true) minimum. Therefore, the system can remain around the higher minimum for some time.

8.4 The Clausius–Clapeyron Law

The slope of the coexistence line in the phase diagram, $\mathrm{d}P/\mathrm{d}T$, can be related to measurable quantities. Along the coexistence line, the Gibbs free energies of the two phases are the same. Therefore, we have two equations that apply at adjacent points on the coexistence line:

$$\begin{aligned} G_{\mathrm{I}}(T, P, N) &= G_{\mathrm{II}}(T, P, N)\,, \\ G_{\mathrm{I}}(T + \Delta T, P + \Delta P, N) &= G_{\mathrm{II}}(T + \Delta T, P + \Delta P, N)\,, \end{aligned} \tag{8.8}$$

where I and II are labels used to distinguish the two phases. We expand both sides of the second equation with respect to ΔT and ΔP to the lowest order and, using the relation in the first line, obtain

$$\Delta T \left(\frac{\partial G_{\mathrm{I}}}{\partial T}\right)_{P,N} + \Delta P \left(\frac{\partial G_{\mathrm{I}}}{\partial P}\right)_{T,N} = \Delta T \left(\frac{\partial G_{\mathrm{II}}}{\partial T}\right)_{P,N} + \Delta P \left(\frac{\partial G_{\mathrm{II}}}{\partial P}\right)_{T,N} . \tag{8.9}$$

Since

$$\left(\frac{\partial G_{\mathrm{I}}}{\partial T}\right)_{P,N} = -S_{\mathrm{I}}\,, \tag{8.10}$$

$$\left(\frac{\partial G_{\mathrm{I}}}{\partial P}\right)_{T,N} = V_{\mathrm{I}}\,, \tag{8.11}$$

and so on, we obtain

$$\Delta T\, \Delta S = \Delta P\, \Delta V\,, \tag{8.12}$$

where $\Delta S \equiv S_{\mathrm{II}} - S_{\mathrm{I}}$ and $\Delta V \equiv V_{\mathrm{II}} - V_{\mathrm{I}}$. Since $T\,\Delta S = Q_{\mathrm{L}}$ is the latent heat of the transition,

$$\frac{\mathrm{d}P}{\mathrm{d}T} = \lim_{\Delta T \to 0} \frac{\Delta P}{\Delta T} = \frac{Q_{\mathrm{L}}}{T\,\Delta V} . \tag{8.13}$$

Thus the volume change and the latent heat determine the slope of the coexistence line.

In the case of a gas–liquid transition, where phase I is the liquid and phase II is the gas, $\Delta V > 0$ and $\Delta S > 0$. Therefore, the slope is positive. As we increase the temperature, the saturated vapor pressure increases. In the case of water, the vapor pressure reaches atmospheric pressure at 100 °C, and water begins to boil at that temperature at sea level.

Usually, the entropy and volume of the liquid phase are larger than those of the solid phase. Therefore, the slope of the coexistence line is usually also positive in the case of solid–liquid phase transitions. However, there are exceptions. One such exception is the case of water. As we know, ice floats on water: the volume of a sample of ice is larger than that of a sample of water of the same mass. On the other hand, ice has a lower entropy, which is evident because we need heat to melt ice. Therefore, the slope of the phase boundary between ice and liquid water is negative. Ice melts when pressure is applied.[4] A rough value for the slope can be estimated from our daily experience. To make iced coffee, we first almost fill the glass with ice at $0\,^\circ\mathrm{C}$. Since there is space between the blocks of ice, we can pour nearly the same amount of boiling coffee into the glass. Then most of the ice melts, and we get cold coffee. This tells us that the latent heat, the quantity of heat required to melt a sample of ice, is nearly the same as the quantity of heat required to heat the same amount of water from $0\,^\circ\mathrm{C}$ to $100\,^\circ\mathrm{C}$. The latter quantity of heat is about $100\,\mathrm{cal\,g^{-1}}$, as this can be related to the definition of the calorie, which is equal to 4.18 J in SI units. The volume change can be guessed from the way ice floats on water. About 10% of the ice is above the surface of the water, and so the volume change is about 10%, or $10^{-7}\ \mathrm{m^3\,g^{-1}}$. Using these values, we can obtain the estimate

$$\frac{\mathrm{d}P}{\mathrm{d}T} = -\frac{418\,\mathrm{J}}{273\,\mathrm{K} \times 10^{-7}\,\mathrm{m^3}} = 1.5 \times 10^7\,\mathrm{Pa\,K^{-1}}\,. \tag{8.14}$$

Thus the freezing temperature decreases by 0.007 K when a pressure of 1 atm is applied.[5]

Another exception is the coexistence line between liquid and solid ^{3}He, the isotope of helium with mass number 3. Liquid helium-3 remains liquid even at zero temperature when the pressure is below about 3.4 MPa. Above that pressure, it solidifies. The phase diagram is as shown in Fig. 8.5. In the low-temperature portion of the coexistence line, between $T \simeq 1\,\mathrm{mK}$ and $T \simeq 0.316\,\mathrm{K}$, the slope is negative. How is such a behavior possible? The solid phase has a smaller volume. This is evident because the solid phase is on the high-pressure side. This is the normal behavior. Therefore, the entropy must behave abnormally. Namely, the entropy of the solid must be larger than that of the liquid. The larger entropy of the solid cannot originate from the motion

[4] This melting of ice under pressure was once considered to be the main reason why we can enjoy skiing or ice-skating. However, this is not correct. The lowering of the melting temperature under a skater's blade is estimated to be only about 3 K, which is too small to explain the fact that we can skate or ski even when the temperature of the ice or snow is $-30\,^\circ\mathrm{C}$. The true reason is the existence of a thin liquid layer covering the ice, although the reason for the existence of this liquid is not fully understood [4].

[5] The correct values for the latent heat and the molar-volume difference at ambient pressure are $Q_\mathrm{L} = 6.01 \times 10^3\,\mathrm{J\,mol^{-1}}$ and $\Delta V = 1.62 \times 10^{-6}\,\mathrm{m^3\,mol^{-1}}$. Therefore, $\mathrm{d}P/\mathrm{d}T = 1.36 \times 10^7\,\mathrm{Pa\,K^{-1}}$.

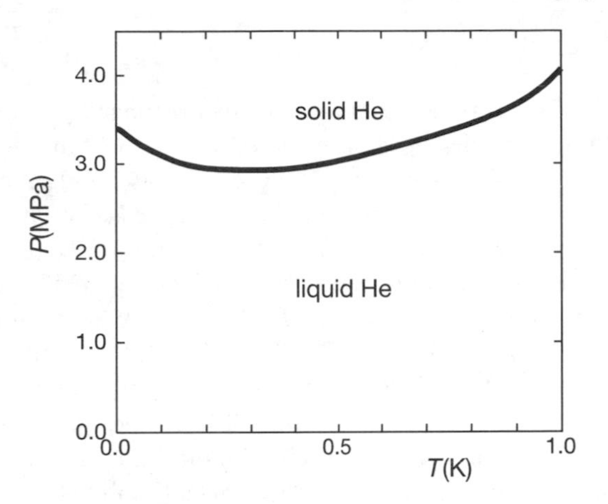

Fig. 8.5. Phase diagram of helium-3 below 1 K. The solid–liquid coexistence line has a minimum pressure of 2.9 MPa at 0.316 K. The phase boundary between the gas phase and the liquid phase is on the low-pressure side of the liquid phase

of the atoms. In fact, it arises from the nuclear magnetic moments. In the solid phase, the atoms are confined around lattice points. If the atoms did not move, the interaction between the nuclear moments would be only the magnetic dipole interaction, which is small because of the small magnetic moment of the nuclei. In reality, cyclic exchanges of atoms are possible even in the solid phase, which leads to an interaction between nuclear moments through quantum mechanical effects. This interaction is larger than the magnetic interaction. However, at the temperatures that we are considering, even though they are low, the interaction is not enough to cause a magnetic phase transition, and the solid He is in a paramagnetic phase. The moment of each atom has two possible orientations, and so for N atoms the contribution to the entropy from the magnetic moments is

$$S_{\mathrm{s}} = k_{\mathrm{B}} N \ln 2 . \tag{8.15}$$

On the other hand, in the liquid phase, the moments cannot be so free. They are restricted so as to make the total moment zero for a reason explained in Chap. 10, and the resulting entropy is therefore much smaller than that for the solid. This difference in the entropy due to the magnetic moments exceeds the difference in the entropy due to the motion of the atoms, and the slope therefore becomes negative. The solid loses this large entropy below the magnetic transition at $T \simeq 0.92\,\mathrm{mK}$. Thus, below this temperature, the slope of the coexistence line returns to the normal behavior.

The fact that the entropy of the solid phase is larger means that when the liquid is compressed so that it is changed into the solid, heat is absorbed. Thus, if it is compressed on the coexistence line, the system moves to lower temperatures. This phenomenon is called Pomeranchuk cooling, and was used to demonstrate the superfluid transition in liquid helium-3 at around 1 mK for the first time in 1972 [5].

8.5 The Critical Point

Let us go back to the phase diagram of an ordinary gas–liquid transition. In Fig. 8.3, the coexistence pressure increases as the temperature increases until the coexistence line terminates at the critical point. Above this point, there is no distinction between the liquid and the gas. What happens here is schematically depicted in Fig. 8.6, where the value of $E + PV - S(E, V)T$ that appears in the integrand of the partition function Y is plotted as a function of V for three pairs of values (T, P) on the coexistence line, where E can be approximated by $U(T, P)$. Therefore, at the minima, this integrand coincides with the Gibbs free energy. As we move to higher temperatures, the two minima corresponding to the liquid phase and the gas phase approach each other, and finally merge into a single minimum at the critical point. Therefore, above the critical point there is no distinction between the two phases. This behavior is similar to what we have seen in the mean-field treatment of a magnetic material described by the Ising model.

The behavior around the critical point can be considered as a kind of phase transition. At higher temperatures, the system is in a phase where there is no distinction between the liquid and the gas. At lower temperatures, it is in a "phase" where there is such a distinction. Again, this is similar to a magnetic phase transition. For a magnetic material, there is a distinction between two directions in the ferromagnetic phase: the direction parallel to the magnetic moment and the direction opposite to it. The distinction disappears above the Curie temperature. These phase transitions are second-order phase transitions, which we discuss in Chap. 9.

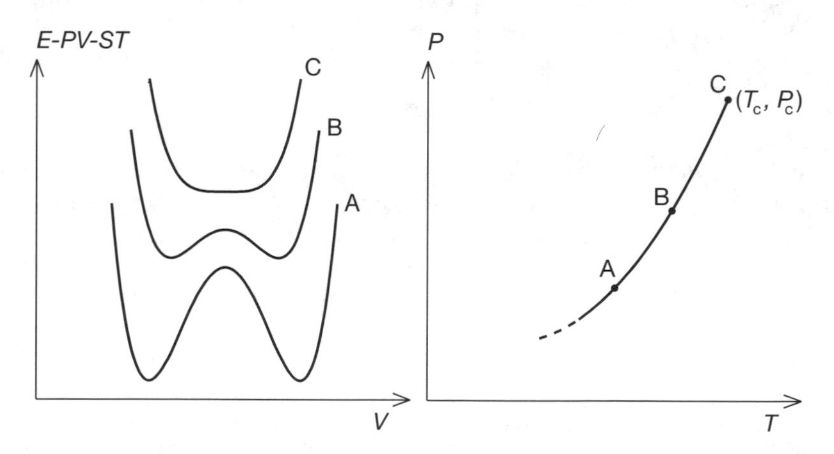

Fig. 8.6. The *left panel* shows the V dependence of the integrand $E+PV-S(E, V)T$ of the partition function near the critical point. The *lines* are displaced vertically by arbitrary amounts. A, B, and C correspond to the points in the phase diagram shown in the *right panel*. Points A and B are on the coexistence line, and point C is the critical point

The reason for the existence of the critical point is that the difference between the liquid and the gas lies only in the density. The difference is qualitative, and can be reduced to zero continuously. On the other hand, a solid differs from a liquid or a gas qualitatively. The atoms or molecules in a solid are on lattice sites, and so there is what is called *long-range order*. That is, from a knowledge of the configuration of the molecules in one part of the solid, the positions of molecules far from that part can be known. This long-range order is the hallmark of a solid.

The existence of the critical point is useful in that it makes it possible to achieve some things that might look difficult. One of these is the production of aerogels. Aerogels are solids of very low density, which are made by replacing water in a gel with air. You might think that they would be easy to make: take a gel and simply evaporate the water from it, and that would be it. However, aerogels cannot be made in this way. Because of the surface tension of the water, the gel is destroyed in the course of the evaporation. This difficulty can be avoided by taking advantage of the existence of a critical point. The gel is put into a pressure cell together with water. The water is placed under a pressure higher than the critical pressure, the temperature is raised higher than the critical temperature, the pressure is reduced below the critical pressure, and the temperature is then reduced, as shown in Fig. 8.7. Now that the water in the gel has been replaced by vapor, air can be introduced safely, and an aerogel can be made. The trick is that during this process there is no coexistence of liquid and gas, and therefore there is no surface and no surface tension.

Water above the critical point is called *supercritical water*. It is useful because it has the properties of both a liquid and a gas. In addition to being used to make aerogels, it is used in various other situations. For example, it

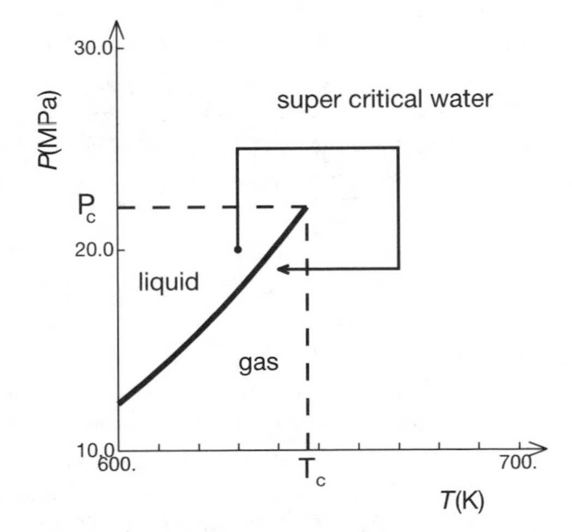

Fig. 8.7. Liquid water can be changed continuously into vapor by going around the critical point

is a good solvent, like ordinary water, but the solute can diffuse rapidly in the solution like gas molecules. Therefore, chemical reactions can be carried out efficiently in supercritical water. It is also powerful for decomposing organic molecules by hydrolysis. Therefore, it can be used to decompose harmful substances such as dioxins and polychlorinated biphenyls (PCBs).

8.6 The van der Waals Gas

The equations of state of real gases deviate from the Boyle–Charles law. The *van der Waals equation* is often used to describe the equations of state of real gases. The equation for one mole of gas contains parameters a and b, which are specific to each gas:

$$\left(P + \frac{a}{V^2}\right)(V - b) = RT\,. \tag{8.16}$$

This equation can also be written in the form

$$P = \frac{RT}{V - b} - \frac{a}{V^2}\,. \tag{8.17}$$

In this equation, a describes the attractive interaction between molecules, and b describes the hard-core part of the repulsive interaction. Because of the attractive interaction between the molecules, a molecule at the boundary of the system is pulled inwards, and therefore the pressure on the wall is reduced. This effect is taken into account by the parameter a, which makes the pressure lower than in the ideal-gas case. Since V^{-2} is proportional to the mean distance between the molecules to the sixth power, it reflects the r^{-6} dependence of the van der Waals potential. The parameter b is easier to understand. Because of the hard-core repulsion, the system cannot have a volume smaller than the volume at which the molecules are closely packed together. This minimum volume is b, and at this volume the pressure diverges. The values of the parameters for typical gases are listed in Table 8.1. The van der Waals equation is interesting because it predicts a critical point and a gas–liquid phase transition.

The isothermal P–V relation predicted by the van der Waals equation is shown in Fig. 8.8. At higher temperatures, the pressure decreases monotonically as the volume increases. However, at low temperatures, the pressure first decreases rapidly, then begins to increase, and decreases again at larger volumes. The temperature T_c that separates these two distinct behaviors can be determined from the equation of state. At higher temperatures, $(\partial P/\partial V)_T$ is always negative, but at lower temperatures it becomes zero at two values of V. Therefore, at $T = T_c$, $(\partial P/\partial V)_T \propto -(V - V_c)^2$ close to the point at which it is zero. From this condition, $T_c = (8/27)a/Rb$ and $V_c = 3b$ can be deduced. (See Exercise 16 at the end of this chapter for a hint about to derive these relations.) This temperature is the critical temperature, and the critical

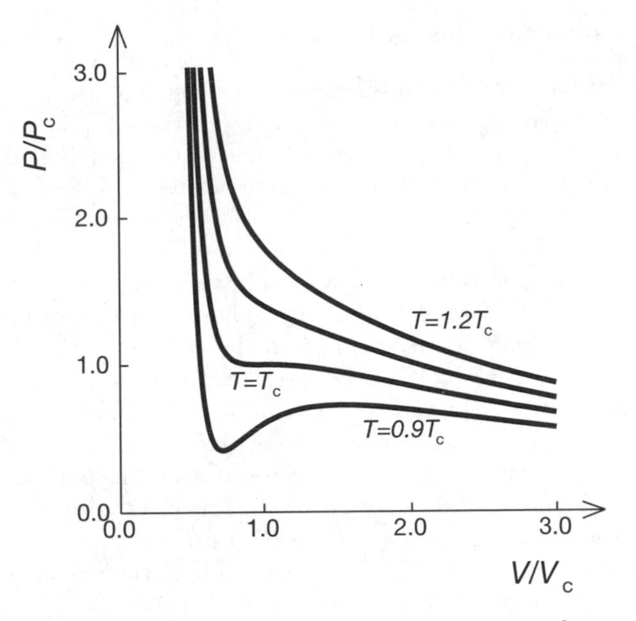

Fig. 8.8. Isotherms of a van der Waals gas. The temperature, volume, and pressure are scaled by $T_c = (8/27)a/Rb$, $V_c = 3b$, and $P_c = (1/27)a/b^2$, respectively

pressure P_c is equal to $(1/27)a/b^2$. If T, V, and P are scaled by these values to obtain $t \equiv T/T_c$, $v \equiv V/V_c$, $p \equiv P/P_c$, the equation of state can be written in a material-independent form,

$$p = \frac{8t}{3v - 1} - \frac{3}{v^2} \,. \tag{8.18}$$

The values of T_c calculated from the parameters in Table 8.1 and the actual values are compared in that table.

Table 8.1. Parameters a and b used to describe typical gases by means of the van der Waals equation. (From Epstein [6]; 1 atm = 1.01325×10^5 Pa.) The critical temperatures predicted from these values, T_c^{calc}, and the actual critical temperatures T_c^{actu} are also shown

Gas	a (atm m^6 mol^{-2})	b (m^3 mol^{-1})	T_c^{calc} (K)	T_c^{actu} (K)
He	0.03415×10^{-6}	23.71×10^{-6}	5.20	5.25
Ne	0.2120×10^{-6}	17.10×10^{-6}	44.8	44.8
H_2	0.2446×10^{-6}	26.61×10^{-6}	33.19	33.25
N_2	1.346×10^{-6}	38.52×10^{-6}	126.1	126.1
O_2	1.361×10^{-6}	32.58×10^{-6}	150.8	154.4
CO_2	3.959×10^{-6}	42.69×10^{-6}	334.8	304.3
H_2O	5.468×10^{-6}	30.52×10^{-6}	646.8	647.3

8.6.1 Coexistence of Gas and Liquid

There is no distinction between the gas and the liquid above T_c. On the other hand, below T_c, part of the isothermal line describes the equation of state of the gas and the other part describes that of the liquid. Let us consider an isotherm at $T < T_c$ and draw a horizontal line at $P = P_{\text{coex}}$, as shown in Fig. 8.9. The pressure $P = P_{\text{coex}}$ has been chosen such that the shaded areas above and below the line have equal areas. This P_{coex} is the coexistence pressure at this temperature. The lowest-volume intersection of this line with the isotherm gives the volume of the liquid phase V_l at the coexistence point, and the highest-volume intersection gives the volume of the gas phase V_g. The part of the isotherm at $V < V_l$ describes the P–V relation for the liquid phase, and the part of the isotherm at $V > V_g$ describes P–V relation for the gas phase. The other parts of the isotherm with a negative slope describe a metastable supercooled or superheated state; the part with a positive slope is unstable and can never be realized. This method of determining the coexistence condition is known as *Maxwell's rule*.

The fact that P_{coex} gives the coexistence pressure can be shown by calculating the difference in the Gibbs free energy between the liquid phase and the gas phase. Since

$$\left(\frac{\partial G}{\partial P}\right)_T = V\,, \tag{8.19}$$

the difference between the free energies at (V_l, P_{coex}) and (V_g, P_{coex}) should be given by the integral along the isotherm:

$$G(V_g, P_{\text{coex}}) - G(V_l, P_{\text{coex}}) = \int_{P_{\text{coex}}}^{P_{\text{coex}}} V\,\mathrm{d}P\,. \tag{8.20}$$

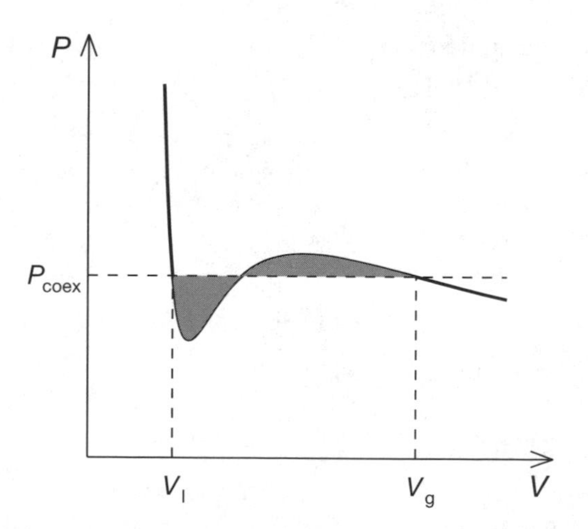

Fig. 8.9. An isotherm below the critical temperature and a method to determine the gas–liquid coexistence pressure

Here the integration is performed from V_l to V_g. We separate this integral into four parts and consider the meaning of each part. The first part is from P_{coex} to P_m, the minimum of the isotherm. This integral gives the negative of the area bounded by two horizontal lines at P_{coex} and P_m, the vertical axis at $V = 0$, and the isotherm. The negative sign arises from the fact that $P_{coex} > P_m$. The next part is the integral from P_m to P_{coex}, namely to the point where the coexistence pressure intersects the isotherm in the unstable region. This integral also gives the area to the left of the isotherm, but this time the sign is positive. Therefore, the sum of the first and the second part gives the shaded area below the coexistence pressure. Similarly, the third and the fourth part give the negative of the shaded area above the coexistence pressure. As a result, the whole integral gives zero when the two shaded regions have the same area, and then the Gibbs free energies have the same value. A similar consideration tells us that when the Gibbs free energies are compared at a pressure higher than P_{coex}, the Gibbs free energy of the liquid phase has a lower value, and when the Gibbs free energies are compared at a pressure lower than P_{coex}, the Gibbs free energy in the gas phase has lower value.

The van der Waals equation of state was written down phenomenologically by considering the attractive and repulsive parts of the interaction between gas molecules. It is not a rigorous equation of state, but it captures the essence of the effects of interaction, and predicts a gas–liquid transition. A Nobel Prize was awarded to van der Waals in 1910 for the discovery of this equation of state.

Exercise 16. When the temperature is fixed at the critical temperature T_c, the equation of state of a van der Waals gas behaves as $P - P_c \propto -(V - V_c)^3$ close to the critical point. Therefore, at (T_c, V_c),

$$\left(\frac{\partial P}{\partial V}\right)_T = 0 \tag{8.21}$$

and

$$\left(\frac{\partial^2 P}{\partial V^2}\right)_T = 0 \tag{8.22}$$

are satisfied. Determine V_c, T_c, and P_c from these equations.

9

Second-Order Phase Transitions

Besides first-order phase transitions, there are other kinds of phase transitions. Here, we consider second-order phase transitions, in which the internal energy changes continuously, and there is no latent heat at the transition. The magnetic phase transition considered in Chap. 7 is an example of such a transition. To describe the transition, we introduce a variable called the *order parameter*, which characterizes one of the phases. As an example of a second-order phase transition, we investigate the two-dimensional Ising model.

9.1 Various Phase Transitions and Order Parameters

A typical example of a second-order phase transition is the magnetic phase transition in the Ising model that we considered in Chap. 7. At temperatures higher than the Curie temperature T_c, the system is in the paramagnetic phase, in which the magnetization is proportional to the magnetic field, whereas at temperatures lower than the Curie temperature, the system is in the ferromagnetic phase, where the magnetization has a nonzero value even without a magnetic field. At the Curie temperature, the magnetization in the absence of a magnetic field changes continuously, and so does the internal energy.

The spins in the Ising model are allowed to point in the $\pm z$ directions only. If we relax this restriction, we can consider other, more realistic models. The XY model is a model in which the spins can point in any direction in the xy plane. The Heisenberg model is a model in which the spins can point in any three-dimensional direction. These models can also undergo a second-order phase transition. The higher-temperature phase is a paramagnetic phase, and the lower-temperature phase is a ferromagnetic phase with a nonzero magnetic moment, which is a vector in these models.

Another example is provided by the critical point of the gas–liquid phase transition described in the previous chapter. Above the critical temperature, there is no distinction between the gas and liquid phases, but below the critical temperature, the densities of these phases are different. The difference in the density develops continuously, like the magnetization, at the critical point, and the internal energy changes continuously. The normal-metal-to-superconductor phase transition and the normal-fluid-to-superfluid phase transition are also examples of second-order phase transitions.

Since the transition is continuous, there should be something which allows us to make a distinction between the phases above and below the transition temperature, otherwise we would not be able to detect the existence of the transition. In many cases the distinction appears in the symmetry of the system. In this case the higher-temperature phase has the higher symmetry, and the lower-temperature phase has the lower symmetry. For example, in the case of the Ising model, the original system has no preference between the up and down directions of each spin. This symmetry is retained in the paramagnetic phase. The up direction and the down direction have equal probabilities of being realized, and so the average magnetic moment vanishes in the absence of a magnetic field. In the ferromagnetic phase, this symmetry is spontaneously broken. The system has a nonzero magnetic moment pointing either in the up direction or in the down direction.

These two possibilities for the magnetization have the same total energy, and so, from the viewpoint of the canonical distribution, should be realized with equal probability. However, in real systems only one of these possibilities is realized. This is because the low-temperature phase is created by the cooperation of the spins. Each spin has a favored direction, because the surrounding spins are ordered. Therefore, most of the spins must change direction simultaneously for the system to move from one choice of direction to the other, which is impossible for a macroscopic system. The creation of a symmetry-broken state is called *spontaneous symmetry breaking.*

Although spontaneous symmetry breaking signals the phase transition in most cases of second-order phase transitions, and it is appropriate to make use of this concept to understand the transition qualitatively, for quantitative discussion of a phase transition it is better to introduce a quantitative variable called the "order parameter". For magnetic systems, the order parameter is the total magnetic moment, which is generally a vector; for the gas–liquid transition, it is the difference in the density; for superfluid helium, it is the wave function of the Bose–Einstein condensate; and for a superconductor, it is the wave function of the Cooper pairs. These order parameters vanish continuously as the temperature approaches the transition temperature from below, and remain zero in the higher-temperature phase.

9.2 Landau Theory

Here we consider the general structure of a second-order phase transition. For simplicity, we consider the case in which the order parameter Ψ is a real scalar.

Every microscopic state that is realized in the system under consideration has some value for this Ψ. For a system in a heat bath, each microscopic state is realized with a probability that is given by the canonical distribution. The value of the order parameter that is obtained at a given temperature is the most probable value of the order parameter at that temperature.

To calculate the probability distribution of the order parameter, we define the density of states $\Omega(E, \Psi)$ for a given value of the order parameter Ψ. Namely,

$$\Omega(E, \Psi)\, \mathrm{d}E\, \mathrm{d}\Psi \tag{9.1}$$

gives the number of microscopic states that have an energy between E and $E + \mathrm{d}E$ and an order parameter between Ψ and $\Psi + \mathrm{d}\Psi$. We then define the partial partition function $z(\Psi)$ by

$$z(\Psi) \equiv \int_0^\infty \mathrm{d}E\, \Omega(E, \Psi) \mathrm{e}^{-\beta E} \,. \tag{9.2}$$

The total partition function Z is given by the integral of $z(\Psi)$:

$$\begin{aligned} Z &= \int_{-\infty}^{\infty} \mathrm{d}\psi \int_0^\infty \mathrm{d}E\, \Omega(E, \Psi) \mathrm{e}^{-\beta E} \\ &= \int_{-\infty}^{\infty} \mathrm{d}\psi\, z(\psi) \,. \end{aligned} \tag{9.3}$$

The probability of a particular value of Ψ being realized is given by

$$\frac{z(\Psi)}{Z} \,. \tag{9.4}$$

To discuss this probability, Landau expressed the partial partition function in terms of a free energy $F_{\mathrm{L}}(T, V, N, \Psi)$, where

$$F_{\mathrm{L}}(T, V, N, \Psi) \equiv -k_{\mathrm{B}} T \ln \left[z(\Psi) \right] \,. \tag{9.5}$$

Landau expected that near the phase transition point the order parameter Ψ would be small, and so the Landau free energy F_{L} can be expanded as a power series in Ψ. There should be no terms containing odd powers of Ψ, since we are considering a system in which states with order parameters $+\Psi$ and $-\Psi$ are equally probable at high temperature, although this symmetry is spontaneously broken below T_{c}. The general form is

$$F_{\mathrm{L}}(T, V, N, \Psi) = F_0(T, V, N) + a\Psi^2 + b\Psi^4 + \cdots \,. \tag{9.6}$$

The most probable, and almost certainly realized, value of the order parameter at (T, V, N) is determined by the minimum of this free energy. Since Ψ is small near the transition point, we can safely neglect the terms of higher order than Ψ^4 in the expansion of (9.6).

This Landau free energy has a minimum at $\Psi = 0$ if the coefficients a and b are positive. Therefore, $a > 0$ and $b > 0$ describes the phase above the critical temperature. On the other hand, if $a < 0$ and $b > 0$, minima occur at

$$\Psi = \pm\sqrt{-\frac{a}{2b}}\,. \tag{9.7}$$

Therefore, this situation describes the ordered phase below the critical temperature, and only one of the two possible choices of the order parameter is realized. It is natural to assume that the coefficient a has the following T dependence near the critical temperature T_c:

$$a(T) = a_0(T - T_\mathrm{c})\,. \tag{9.8}$$

The order parameter then has the following temperature dependence:

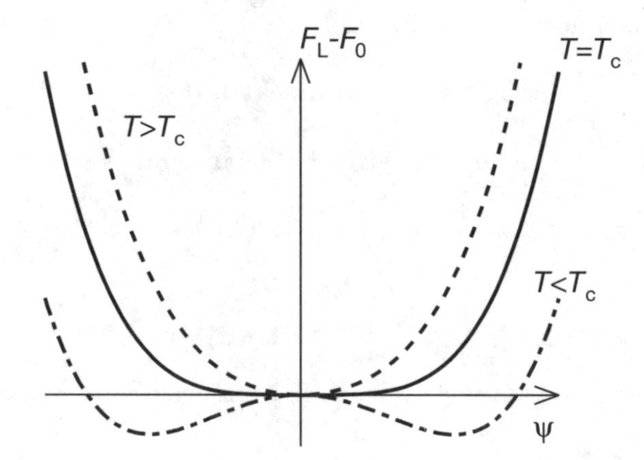

Fig. 9.1. Dependence of $F_\mathrm{L}(T,V,N,\Psi) - F_0(T,V,N)$ on Ψ at $T < T_\mathrm{c}$ (*dash-dotted line*), $T = T_\mathrm{c}$ (*solid line*), and $T > T_\mathrm{c}$ (*dashed line*)

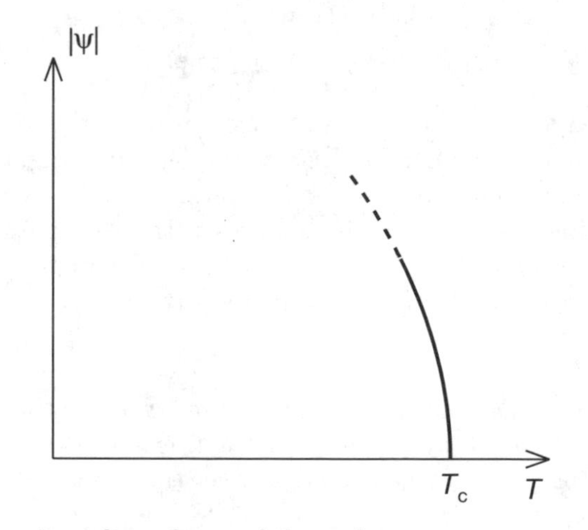

Fig. 9.2. Temperature dependence of the order parameter around T_c. The order parameter is proportional to $\sqrt{T_\mathrm{c} - T}$ for $T < T_\mathrm{c}$

$$\Psi = \begin{cases} 0 & (T \geq T_c)\,, \\ \pm\sqrt{\dfrac{a_0}{2b}(T_c - T)} & (T < T_c)\,. \end{cases} \tag{9.9}$$

The fact that the order parameter is proportional to $\sqrt{T_c - T}$ for $T < T_c$ is the same behavior as in the mean-field result for the Ising model. The behavior of the Landau free energy and the order parameter is depicted in Figs. 9.1 and 9.2.

9.2.1 Free Energy

Now we consider the behavior of various thermodynamic variables around the transition temperature. The total partition function Z, which gives the true free energy $F(T, V, N)$, is obtained from an integral containing $F_L(T, V, N, \Psi)$:

$$Z(T, V, N) = \int \mathrm{d}\Psi\, z(\Psi) = \int \mathrm{d}\Psi\, \mathrm{e}^{-\beta F_L(T,V,N,\Psi)}\,. \tag{9.10}$$

The integrand is expected to be sharply peaked at the most probable value of Ψ for a macroscopic system. Thus, the integration can be done as a Gaussian integral. For $T > T_c$, we obtain

$$Z(T, V, N) = \mathrm{e}^{-\beta F_0(T,V,N)} \int_{-\infty}^{\infty} \mathrm{d}\Psi\, \mathrm{e}^{-\beta a \Psi^2} = \sqrt{\frac{\pi}{\beta a}}\mathrm{e}^{-\beta F_0(T,V,N)} \tag{9.11}$$

and

$$F(T, V, N) = -k_B T \ln Z = F_0(T, V, N) - \frac{1}{2} k_B T \ln\left(\frac{\pi}{\beta a}\right) \simeq F_0(T, V, N)\,. \tag{9.12}$$

The last approximation is justified because F_0 is a macroscopic quantity of order $O(N) \times k_B T$ but the neglected term is of the order of $k_B T$, unless $a \simeq 0$.

For $T < T_c$, only one of the two possibilities for the order parameter is realized. Once one value has been chosen by the system, the broken symmetry does not allow the other value to be realized in the ordered phase. Thus, only microscopic states around the selected order parameter are counted in the partition function. Here we choose $\Psi = \sqrt{-a/2b}$. Around this value, we expand F_L as follows:

$$F_L(T, V, N, \Psi) \simeq F_0(T, V, N) - \frac{a^2}{4b} - 2a\,(\Delta\Psi)^2\,, \tag{9.13}$$

where

$$\Delta\Psi \equiv \Psi - \sqrt{\frac{-a}{2b}}\,. \tag{9.14}$$

A Gaussian integral around this minimum gives the partition function and the free energy for $T < T_c$:

$$
\begin{aligned}
Z(T,V,N) &\simeq \int_{-\infty}^{\infty} \mathrm{d}(\Delta\Psi)\, \exp\left[-\beta\left(F_0(T,V,N) - \frac{a^2}{4b} - 2a\,(\Delta\Psi)^2\right)\right] \\
&= \sqrt{\frac{\pi}{-2a\beta}}\, \exp\left[-\beta\left(F_0(T,V,N) - \frac{a^2}{4b}\right)\right]
\end{aligned}
\tag{9.15}
$$

and

$$
\begin{aligned}
F(T,V,N) &= F_0(T,V,N) - \frac{a^2}{4b} - \frac{1}{2}k_{\mathrm{B}}T \ln\left(\frac{\pi}{-2a\beta}\right) \\
&\simeq F_0(T,V,N) - \frac{a_0^2}{4b}(T_{\mathrm{c}} - T)^2 .
\end{aligned}
\tag{9.16}
$$

The last approximation is valid except close to the transition temperature, where $a \simeq 0$. In this approximation, the free energy changes continuously through the critical temperature.

9.2.2 Entropy, Internal Energy, and Heat Capacity

Here we consider how the thermodynamic variables behave in the ordered phase compared with the normal phase that exists when $T > T_{\mathrm{c}}$. The entropy S is given by the derivative of F. For $T > T_{\mathrm{c}}$,

$$
\begin{aligned}
S(T,V,N) &= -\left(\frac{\partial F(T,V,N)}{\partial T}\right)_{V,N} = -\left(\frac{\partial F_0(T,V,N)}{\partial T}\right)_{V,N} \\
&\equiv S_0(T,V,N) .
\end{aligned}
\tag{9.17}
$$

The internal energy U is given by

$$
\begin{aligned}
U(T,V,N) &= F(T,V,N) + S(T,V,N)T \\
&= F_0(T,V,N) + S_0(T,V,N)T .
\end{aligned}
\tag{9.18}
$$

The constant-volume heat capacity C is given by

$$
\begin{aligned}
C(T,V,N) &= \left(\frac{\partial U}{\partial T}\right)_{V,N} = T\left(\frac{\partial S(T,V,N)}{\partial T}\right)_{V,N} \\
&\equiv C_0(T,V,N) .
\end{aligned}
\tag{9.19}
$$

The concrete behavior of these quantities above T_{c} depends on the actual system.

Next we calculate these variables for the low-temperature phase. The entropy and the internal energy have the following forms below T_{c}:

$$
\begin{aligned}
S(T,V,N) &= -\left(\frac{\partial F}{\partial T}\right)_{V,N} = -\left(\frac{\partial F_0}{\partial T}\right)_{V,N} - \frac{a_0^2}{2b}(T_{\mathrm{c}} - T) \\
&= S_0(T,V,N) - \frac{a_0^2}{2b}(T_{\mathrm{c}} - T)
\end{aligned}
\tag{9.20}
$$

and

$$
\begin{aligned}
U(T,V,N) &= F(T,V,N) + S(T,V,N)T \\
&= F_0(T,V,N) + S_0(T,V,N)T + \frac{a_0^2}{4b}\left(T^2 - T_c^2\right) .
\end{aligned}
\tag{9.21}
$$

The entropy and the internal energy are continuous at the transition temperature.

The heat capacity C is not continuous. It has the following form for $T < T_c$:

$$
\begin{aligned}
C(T,V,N) = T\left(\frac{\partial S}{\partial T}\right)_{V,N} &= T\left(\frac{\partial S_0}{\partial T}\right)_{V,N} + \frac{a_0^2}{2b}T \\
&= C_0(T,V,N) + \frac{a_0^2}{2b}T .
\end{aligned}
\tag{9.22}
$$

The behavior of C is depicted in Fig. 9.3; it has a discontinuity of magnitude $(a_0^2/2b)T_c$ at the transition temperature.

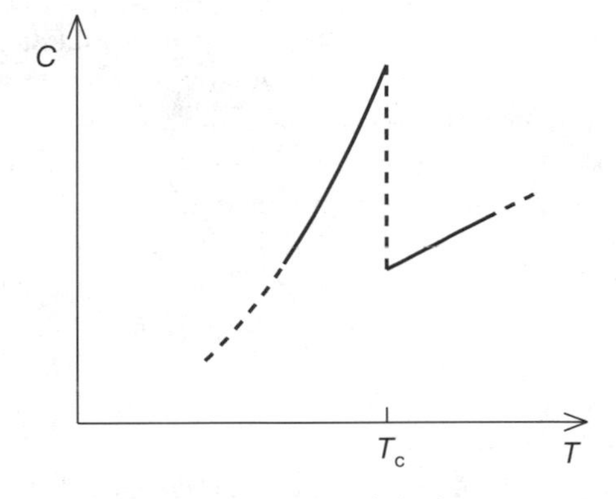

Fig. 9.3. Temperature dependence of the heat capacity around T_c

9.2.3 Critical Phenomena

The Landau theory predicts the general behavior of the thermodynamic variables around the transition temperature, as described above. In the present section we have restricted the discussion to the case in which the order parameter is a real scalar. The theory is easily generalized to cases where the order parameter is a vector, as in the case of the Heisenberg model of a ferromagnet, or where it is a complex scalar, as in the case of a superfluid, and so on. So this theory is quite powerful as a means of helping us to understand second-order phase transitions in general. It is known that, for various systems, the Landau theory gives the same behavior as the mean-field theory around the critical temperature.

However, the Landau theory fails to describe systems when the temperature is very close to the transition temperature. The reason is easily understood. As we have remarked, close to the transition point, the coefficient $a(T)$ becomes small. Thus, it is no longer appropriate to take into account only the quadratic term and to neglect the higher-order terms in the expansion of the Landau free energy in terms of the order parameter. Neglect of the higher-order terms leads to a divergence of the free energy at $a = 0$, which does not occur in reality. Around the transition temperature, we need to treat the system more carefully.

The investigation of the behavior of systems close to the transition temperature is an active area of research in statistical physics; it is referred to as the study of critical phenomena. The main theme of such investigations is the determination of *critical exponents*. That is, the most dominant temperature dependences of various variables around T_c are assumed to have a form $|T - T_\mathrm{c}|^x$, and the exponent x is called the critical exponent. Critical exponents are denoted by Greek letters: for example, the critical exponent for the heat capacity is represented by α, and that for the order parameter is represented by β. In the Landau theory, the order parameter has the behavior $\Psi \propto (T_\mathrm{c} - T)^\beta$, where $\beta = 1/2$, and so the critical exponent β for the order parameter is $1/2$. In the Landau theory, the heat capacity shows a discontinuous change and does not show a power-law behavior, and so $\alpha = 0$. Since the Landau theory fails at the transition temperature, the critical exponents obtained by this theory do not agree with experiment in general. Therefore, critical phenomena have been investigated intensively. Going beyond the Landau theory is beyond the scope of this book, however. In the next section we give the exact free energy for the two-dimensional Ising model, and see how the system behaves at the transition point, as an example of a critical phenomenon.

9.3 The Two-Dimensional Ising Model

We consider a two-dimensional Ising model on a square lattice. In this model, the spins are placed at lattice points as shown in Fig. 9.4. The ith spin σ_i can take a value of either 1 or -1. The energy of a configuration $\{\sigma_i\}$ is given by

$$E = -\sum_{i,j} J_{i,j}\sigma_i\sigma_j\,, \tag{9.23}$$

where the interaction energy $J_{i,j}$ is either J or J', depending on whether the ith and jth spins are nearest neighbors in the horizontal direction or in the vertical direction, respectively. Onsager calculated the free energy for this model analytically in 1944 [7]. The result is

$$\begin{aligned} F = -k_\mathrm{B}TN\Bigg\{&\frac{1}{2}\ln\left[4\cosh(2\beta J)\cosh(2\beta J')\right] \\ &+\frac{1}{2\pi^2}\int_0^\pi \mathrm{d}\omega \int_0^\pi \mathrm{d}\omega' \ln\left(1 - 2\kappa\cos\omega - 2\kappa'\cos\omega'\right)\Bigg\}\,, \end{aligned} \tag{9.24}$$

where

$$2\kappa \equiv \frac{\tanh(2\beta J)}{\cosh(2\beta J')}$$

and

$$2\kappa' \equiv \frac{\tanh(2\beta J')}{\cosh(2\beta J)} . \tag{9.25}$$

This model reduces to a one-dimensional model when J' is zero. In this case $2\kappa = \tanh(2\beta J)$, $2\kappa' = 0$, and $\cosh(2\beta J') = 1$. The free energy then reduces to

$$\begin{aligned} F &= -k_{\mathrm{B}}TN \left\{ \frac{1}{2} \ln\left[4\cosh(2\beta J)\right] + \frac{1}{2\pi^2} \int_0^{\pi} \mathrm{d}\omega \int_0^{\pi} \mathrm{d}\omega' \, \ln\left(1 - 2\kappa\cos\omega\right) \right\} \\ &= -k_{\mathrm{B}}TN \left\{ \frac{1}{2} \ln\left[4\cosh(2\beta J)\right] + \frac{1}{2\pi} \int_0^{\pi} \mathrm{d}\omega \, \ln\left(1 - 2\kappa\cos\omega\right) \right\} . \end{aligned} \tag{9.26}$$

The integral here is known to have the following value:

$$\frac{1}{2\pi} \int_0^{\pi} \ln\left(1 - 2\kappa\cos\omega\right) \mathrm{d}\omega = \frac{1}{2} \ln\left(\frac{1}{2} + \frac{1}{2}\sqrt{1-4\kappa^2}\right) . \tag{9.27}$$

Since $2\kappa = \tanh 2\beta J$, $\sqrt{1-4\kappa^2} = 1/\cosh(2\beta J)$. We then obtain

$$\begin{aligned} F &= -k_{\mathrm{B}}TN \left\{ \frac{1}{2} \ln\left[4\cosh(2\beta J)\left(\frac{1}{2} + \frac{1}{2}\frac{1}{\cosh(2\beta J)}\right)\right] \right\} \\ &= -k_{\mathrm{B}}TN \left[\frac{1}{2} \ln\left[2\cosh(2\beta J) + 2\right] \right] \\ &= -k_{\mathrm{B}}TN \ln\left[2\cosh(\beta J)\right] . \end{aligned} \tag{9.28}$$

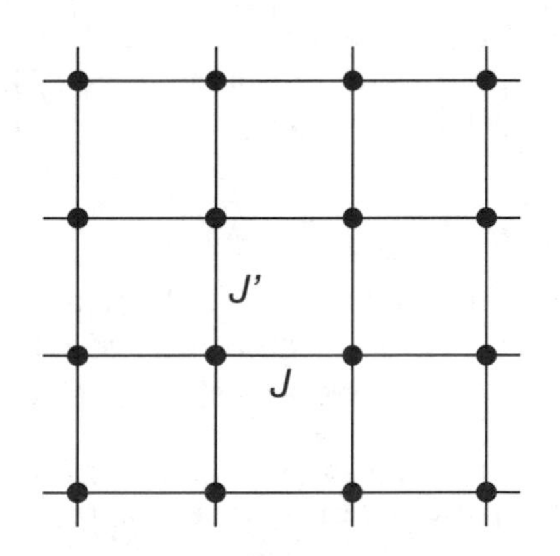

Fig. 9.4. Ising model on a square lattice. Each lattice point (*solid circle*) has a spin σ which can take a value of either 1 or -1. The spins interact with their nearest-neighbor spins. The interaction energy is $-J\sigma_i\sigma_j$ if the ith and jth spins are nearest neighbors in the *horizontal direction*, and $-J'\sigma_i\sigma_j$ if the ith and jth spins are nearest neighbors in the *vertical direction*

The final result agrees with that obtained in Chap. 7.

Now we return to the main theme of this section, the square-lattice Ising model. For a square lattice, $J = J'$ and $2\kappa = 2\kappa' = \tanh(2\beta J)/\cosh(2\beta J)$. The free energy can be transformed as follows:

$$
\begin{aligned}
F &= -k_{\mathrm{B}}TN\bigg(\ln\left[2\cosh(2\beta J)\right] \\
&\quad + \frac{1}{2\pi^2}\int_0^{\pi}\mathrm{d}\omega\int_0^{\pi}\mathrm{d}\omega'\,\ln\left(1-2\kappa\cos\omega-2\kappa\cos\omega'\right)\bigg) \\
&= -k_{\mathrm{B}}TN\bigg(\ln\left[2\cosh(2\beta J)\right] \\
&\quad + \frac{1}{2\pi^2}\int_0^{\pi}\mathrm{d}\omega\int_0^{\pi}\mathrm{d}\omega'\,\ln\left[1-4\kappa\cos(\omega+\omega')\cos(\omega-\omega')\right]\bigg) \\
&= -k_{\mathrm{B}}TN\bigg(\ln\left[2\cosh(2\beta J)\right] \\
&\quad + \frac{1}{2\pi^2}\int_0^{\pi}\mathrm{d}\omega_1\int_0^{\pi}\mathrm{d}\omega_2\,\ln\left[1-4\kappa\cos\omega_1\cos\omega_2\right]\bigg) \\
&= -k_{\mathrm{B}}TN\bigg(\ln\left[2\cosh(2\beta J)\right] \\
&\quad + \frac{1}{2\pi}\int_0^{\pi}\mathrm{d}\omega\,\ln\left[\frac{1}{2}+\frac{1}{2}\sqrt{1-(4\kappa\cos\omega)^2}\right]\bigg) .
\end{aligned}
\tag{9.29}
$$

We have used the formula (9.27) to obtain the final result. In this expression for the free energy, we notice that if κ is larger than 1/4, the square root has a negative argument around $\omega \simeq 0$ and $\omega \simeq \pi$. However, κ can be expressed in the form

$$
\kappa = \frac{1}{2}\,\frac{\sinh(2\beta J)}{1+\sinh^2(2\beta J)} , \tag{9.30}
$$

which has a maximum at $\sinh(2\beta J) = 1$, the maximum value being 1/4. Therefore, the free energy behaves normally over the whole temperature range, although we may anticipate that something special will happen at $\sinh(2\beta J) = 1$.

Our expectation of something special at $\sinh(2\beta J) = 1$ is confirmed when we calculate the heat capacity C. First, we calculate the internal energy

$$
U = -\frac{\partial \ln Z}{\partial \beta} = -NJ\coth(2\beta J)\left[1+\frac{2}{\pi}\kappa_1 K_1\right] , \tag{9.31}
$$

where

$$
\kappa_1 = 2\tanh^2(2\beta J) - 1 \tag{9.32}
$$

and K_1 is the complete elliptic integral of the first kind, defined as follows:[1]

$$K_1 \equiv K(4\kappa) = \int_0^{\pi/2} \frac{\mathrm{d}\phi}{\sqrt{1-(4\kappa\sin\phi)^2}} . \tag{9.33}$$

The heat capacity is then obtained as follows:

$$\begin{aligned} C &= \frac{\partial U}{\partial T} \\ &= Nk_\mathrm{B}\left[\beta J \coth(2\beta J)\right]^2 \frac{2}{\pi}\left[2K_1 - 2E_1 - (1-\kappa_1)\left(\frac{\pi}{2}+\kappa_1 K_1\right)\right] . \end{aligned} \tag{9.34}$$

Here E_1 is the complete elliptic integral of the second kind, defined by

$$E_1 \equiv E(4\kappa) = \int_0^{\pi/2} \sqrt{1-(4\kappa\sin\phi)^2}\,\mathrm{d}\phi . \tag{9.35}$$

Let us write the temperature at which $\sinh(2\beta J)$ becomes 1 as T_c. Since $\sinh(0.881374) = 1$, $T_\mathrm{c} = 2.269185 J/k_\mathrm{B}$. At this temperature, $4\kappa = 1$, and the complete elliptic integral K_1 diverges:

$$K(4\kappa) \simeq \frac{1}{2}\ln\left(\frac{1}{1-4\kappa}\right) \simeq -\ln\left(\frac{2J}{k_\mathrm{B}T_\mathrm{c}^2}|T-T_\mathrm{c}|\right) . \tag{9.36}$$

T_c is the transition temperature, i.e. the Curie temperature, for this model. At T_c, $\kappa_1 = 1$ and $\coth(2\beta_\mathrm{c}J) = \sqrt{2}$. Thus, the heat capacity diverges around this temperature logarithmically:

$$C \simeq -\frac{2}{\pi}Nk_\mathrm{B}\left(\frac{J}{k_\mathrm{B}T_\mathrm{c}}\right)^2 \ln|T-T_\mathrm{c}| . \tag{9.37}$$

The temperature dependence of the heat capacity is shown in Fig. 9.5. This dependence is quite different from the that in the Landau theory, although the critical exponent α in this case is equal to 0 also.[2]

Since $C = T\,\partial S/\partial T$, the peak of C indicates a temperature at which a large increase in the entropy occurs. It is instructive to plot the entropy to see how it increases from $S = 0$ at $T = 0$ to $S = Nk_\mathrm{B}\ln 2$ as $T \to \infty$. It can be obtained from U and F using $S = (U-F)/T$, and is shown in Fig. 9.6.

An analytical expression for the spontaneous magnetization below the Curie temperature has been obtained by Yang [9]. This expression is as follows:

$$M = M_\mathrm{max}\left(1 - \frac{1}{\sinh^4(2\beta J)}\right)^{1/8} . \tag{9.38}$$

[1] For more information about the complete elliptic integrals, see [8], for example.

[2] The divergence of $\ln|T-T_\mathrm{c}|$ at $T \simeq T_\mathrm{c}$ is weaker than $|T-T_\mathrm{c}|^\alpha$ for any $\alpha > 0$. Thus the exponent α of the logarithmic divergence is 0.

Its behavior just below the Curie temperature is

$$M \propto (T_c - T)^{1/8} . \quad (9.39)$$

Therefore, the critical exponent β is $1/8$. The temperature dependence of the magnetization is shown in Fig. 9.7.

As we have seen, the critical exponents for the Ising model differ from those for the Landau theory. There are many models which show a second-order phase transition. Even though an exact calculation of the partition function has not been done for most of these models, the critical exponents have been investigated by various methods, including numerical methods. It has

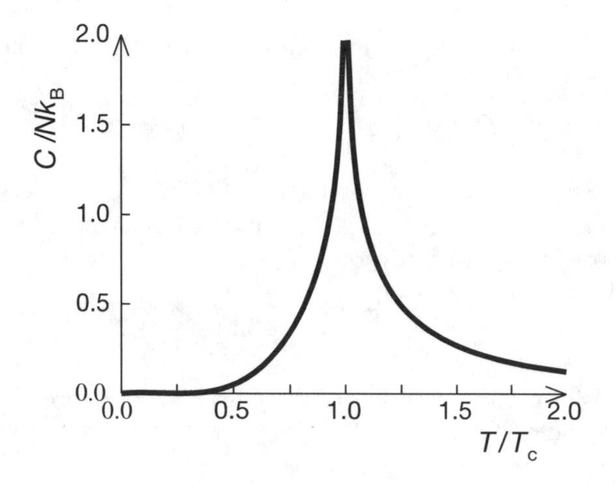

Fig. 9.5. Temperature dependence of the heat capacity for the square-lattice Ising model. The heat capacity diverges logarithmically at the transition temperature $T_c = 2.269185 J/k_B$

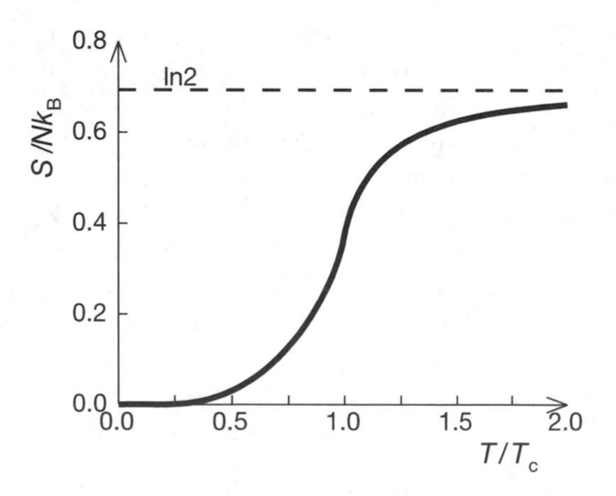

Fig. 9.6. Temperature dependence of the entropy for the square-lattice Ising model. The entropy tends to $Nk_B \ln 2$ as $T \to \infty$

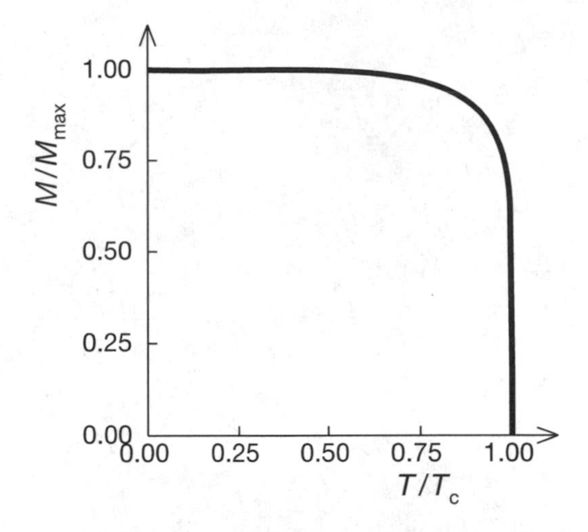

Fig. 9.7. Temperature dependence of the magnetization for the square-lattice Ising model

been shown that these models can be classified into several groups, known as universality classes. Within the same universality class, the critical exponents are the same, and the models behave similarly around the transition temperature. The detailed theory of and explanation for these universality classes are beyond the scope of this book. For those wishing to study critical phenomena further, the books by Stanley [10] and Ma [11] are recommended.

10

Dense Gases – Ideal Gases at Low Temperature

We considered the statistical physics of an ideal gas in Chap. 4. There, we introduced a factor $1/N!$ to reduce the phase space of N molecules, since the atoms or molecules are indistinguishable and an interchange of molecules gives the same microscopic state. However, this procedure is not exact, and this becomes important when the density of the gas becomes sufficiently high. In this chapter we consider what happens in such a system. We shall see that a gas can be classified as either a Fermi gas or a Bose gas depending on the properties of its atoms or molecules.

10.1 The Phase Space for N Identical Particles

In Chap. 4, we introduced the phase space for N particles,[1] which is a $6N$-dimensional space spanned by the three-dimensional space coordinates and three-dimensional momenta of all N particles. The volume of this phase space is reduced by a factor $1/N!$, since any interchange of particles gives the same microscopic state. The number of microscopic states is obtained by dividing this reduced volume by h^{3N}. This reduction of the phase space is almost exactly correct at high temperatures, as we shall show later, but becomes incorrect at low temperatures.

We shall explain the reason for this incorrectness using a very simple system, namely a two-particle system in which each particle has only two possible states, a and b. If the particles are distinguishable, there are four microscopic states: (a,a), (a,b), (b,a), and (b,b), where (a,b) means that the first particle is in state a and the second particle is in state b, and so on. What we did in Chap. 4 was to divide this number of microscopic states, four, by $2! = 2$. Thus, we consider the number of states as two when the particles are indistinguishable. However, of these four states, only two of them, namely (a,b)

[1] In this chapter, we use the word "particles" to represent both atoms and molecules. Later, we shall consider the conduction electrons in a metal as a gas of electrons. Therefore, "particles" is more appropriate than "atoms" or "molecules".

and (b, a), are the same state when the particles are indistinguishable. Hence, the number of microscopic states is not two but three. The discrepancy arises from the fact that there are states in which two particles occupy the same state a or b. An interchange of the particles in (a, b) makes an apparently different state (b, a), but an interchange of the particles in (a, a) or (b, b) does not make a different state. Therefore, division by 2! is not correct in this case, in which double occupancy of the same state by two particles is allowed.

Now let us consider what happens if the states that lead to the discrepancy, those with double occupancy, are not allowed. In this case there is only one allowed state, $(a, b) = (b, a)$, and so the number of microscopic states is not two but one. So even if the origin of the discrepancy is removed, the division by 2! is still not correct. Our present task is to construct a scheme to count the number of microscopic states correctly.

For that purpose, we must have a rule that states whether we allow double occupancy of a state or not. Of course, this rule is determined by nature. It turns out that both possibilities occur. Particles have a characteristic called their "statistics" that determines which possibility. Some particles are called *fermions*, and obey Fermi statistics. For these particles, it is not allowed for a single state to be occupied by more than one particle. That is, for each state, there is only the possibility that it is not occupied or that it is occupied by one particle. This rule for fermions is called the *Pauli exclusion principle*. The remaining particles are classified as *bosons*, and obey Bose statistics. For these particles, any number of particles can occupy the same state.

There is a relation between the spin of a particle and its statistics. A particle with a half-integer spin is a fermion. The electron, proton, and neutron have a spin of one-half, and so they are fermions. On the other hand, a particle with an integer spin is a boson. Electromagnetic radiation can be considered as a collection of particles, called photons. The photon has a spin of 1, and so it is a boson. Lattice vibrations, which contribute to the specific heat of a solid, can also be considered as a collection of particles, called phonons. The phonon has no spin, and so it is a boson. The spin of an atom or molecule is determined by the spins of its constituent particles. An atom or molecule composed of an even number of fermions has an integer spin, and so is a boson, whereas an atom or molecule with an odd number of fermions has a half-integer spin, and so is a fermion. Most helium atoms have a mass number of 4 (^{4}He): each atom of this isotope is composed of two protons, two neutrons, and two electrons. Thus it is a boson. On the other hand, another isotope of helium, ^{3}He, has only one neutron, and so is a fermion.

All particles can be classified as either bosons or fermions. How do we count microscopic states for these particles and how do we justify what we did in Chap. 4? To answer this question, let us consider the general case of a system of N particles, each of which can take one of M states, where

$M \gg N$. If the particles are fermions of the same species, the number of microscopic states is the number of ways to choose N states from M states, ${}_M C_N$. The chosen states are those which accommodate a fermion. In the limit of large $M \gg N \gg 1$, the number of microscopic states can be approximated by

$$ {}_M C_N = \frac{M!}{N!(M-N)!} \simeq \frac{M^N}{N!} \, . \tag{10.1} $$

On the other hand, if the particles are bosons of the same species, the number of microscopic states is the number of ways of distributing N particles among M states without restriction. This is the same as the number of ways to give N yen to M people, which we considered in Chap. 1, and is given by ${}_{M+N-1}C_N$. It can be approximated as follows when $M \gg N \gg 1$:

$$ {}_{M+N-1}C_N = \frac{(M+N-1)!}{N!(M-1)!} \simeq \frac{M^N}{N!} \, . \tag{10.2} $$

These limiting values are the same, and are also equal to the value which was used to count the number of microscopic states in Chap. 4. There, the particles were treated independently; for each particle there are M possibilities and so the total number of ways is M^N, and division by $N!$ gives the same result as above. Therefore, the result in Chap. 4 is justified when $M \gg N$. A gas at normal temperature and pressure satisfies this condition. We shall see this later. On the other hand, this condition is not satisfied at low temperature. Of course, at low temperature the attractive interaction causes the gas to become a liquid, as we saw in Chap. 8. However, there are several systems which can be considered as a gas of particles even at zero temperature. Electrons in metals, liquid helium, and alkali metal vapors are some such systems. We begin our consideration of models for these systems in the next section.

10.2 The Grand Canonical Distribution

Counting the number of microscopic states or calculating the partition function for fermions or bosons is not easy when the total number of particles is fixed. It becomes much easier if we allow fluctuations in the number. Thus, in this section, we consider a system (system I) in a heat bath (system II), with which system I can exchange energy and particles. The heat bath has a fixed temperature T and a fixed chemical potential μ. We shall first describe our general scheme for calculating the partition function and free energy in this case.

First, we consider the probability that system I is in a microscopic state with an energy E_{I} and a number of particles N_{I}. To determine this probability, we consider the total system, i.e. the system plus the heat bath, by use of the microcanonical distribution just as we did in Chap. 3. In this case, the total energy E_{t} and the total number of particles N_{t} are fixed. The probability

$f(E_\mathrm{I}, N_\mathrm{I})$ is proportional to the number of microscopic states of the heat bath:

$$
\begin{aligned}
f(E_\mathrm{I}, N_\mathrm{I}) &\propto \frac{\Omega_\mathrm{II}(E_\mathrm{t} - E_\mathrm{I}, N_\mathrm{t} - N_\mathrm{I})}{\Omega_\mathrm{II}(E_\mathrm{t}, N_\mathrm{t})} \\
&= \exp\left[\frac{1}{k_\mathrm{B}}\left\{S_\mathrm{II}(E_\mathrm{t} - E_\mathrm{I}, N_\mathrm{t} - N_\mathrm{I}) - S_\mathrm{II}(E_\mathrm{t}, N_\mathrm{t})\right\}\right] \\
&\simeq \exp\left[\frac{1}{k_\mathrm{B}}\left\{-E_\mathrm{I}\frac{\partial S_\mathrm{II}}{\partial E} - N_\mathrm{I}\frac{\partial S_\mathrm{II}}{\partial N}\right\}\right] \\
&= \exp\left[-\frac{1}{k_\mathrm{B}T}(E_\mathrm{I} - \mu N_\mathrm{I})\right] . \qquad (10.3)
\end{aligned}
$$

That is, the chemical potential μ of the heat bath controls the probability and hence controls the number of molecules in system I. The corresponding distribution is called the grand canonical distribution.

The normalized probability is

$$
f(E, N) = \frac{1}{\Xi}\exp\left[-\frac{1}{k_\mathrm{B}T}(E - \mu N)\right] . \qquad (10.4)
$$

The denominator of the normalization coefficient,

$$
\begin{aligned}
\Xi(T, \mu) &\equiv \sum_{N=0}^{\infty}\int_0^{\infty} \exp\left[-\frac{(E - \mu N)}{k_\mathrm{B}T}\right]\Omega_\mathrm{I}(E, N)\,\mathrm{d}E \\
&= \sum_{N=0}^{\infty}\int_0^{\infty} \exp\left[-\frac{[E - \mu N - S_\mathrm{I}(E, N)T]}{k_\mathrm{B}T}\right]\mathrm{d}E , \qquad (10.5)
\end{aligned}
$$

is called the *grand partition function.* If system I is macroscopic, the integrand is sharply peaked around the minimum of the argument of the exponential function in the last line, and the system should almost always be found at the values $E = E^*$ and $N = N^*$ for which the minimum is realized. Furthermore, since system I and the heat bath (system II) are in thermal equilibrium, the temperature T and the chemical potential μ are also those of system I.

The free energy associated with this grand partition function can be written as

$$
J = -k_\mathrm{B}T\ln\Xi(T, \mu) . \qquad (10.6)
$$

The value of J can be formally evaluated by expanding the integrand around the maximum at (E^*, N^*) up to second order in the deviation, neglecting terms of order one with respect to terms of order N. The result is

$$
J = E^* - S(E^*, N^*)T - \mu N^* . \qquad (10.7)
$$

Since $E^* = U$ is the internal energy and $\mu N^* = G = U - ST + PV$ is the Gibbs free energy, we arrive at the conclusion that

$$
J = -PV . \qquad (10.8)
$$

10.3 Ideal Fermi Gases and Ideal Bose Gases

10.3.1 Occupation Number Representation

Now we shall apply this scheme to an ideal gas of fermions or bosons. Because we are considering an ideal gas, we can treat each particle independently, except for the limitation due to the Pauli exclusion principle. Therefore, we can define single-particle states for the system. These are states in which individual particles can be accommodated, with energies specific to each state.[2] To describe the microscopic state of a system of many particles, we have considered above a method in which we specify the one-particle states that are occupied by particles. This description tells us which particle is in which single-particle state. However, the same many-particle state can be described by specifying the number of particles accommodated in each single-particle state. That is, we make a table of single-particle states and write down the number of particles in each of these states. This method is called the *occupation number representation*, and is very suitable for the description of fermions and bosons.

Fermion Gases

In this occupation number representation, we can consider each single-particle state, with energy E_i $(i = 1, 2, 3, \cdots, \infty)$, as system I. For a fermion gas, such a state can contain only one particle at most. Therefore, the grand partition function for this single-particle state is

$$\Xi_i = \mathrm{e}^0 + \mathrm{e}^{-\beta(E_i - \mu)} , \tag{10.9}$$

where $\beta = 1/k_{\mathrm{B}}T$ and μ have values determined by the heat bath. The probability that this state contains no particle is

$$\frac{\mathrm{e}^0}{\Xi_i} = \frac{1}{1 + \mathrm{e}^{-\beta(E_i - \mu)}} , \tag{10.10}$$

and the probability that it contains one particle is

$$\frac{\mathrm{e}^{-\beta(E_i - \mu)}}{\Xi_i} = \frac{1}{\mathrm{e}^{\beta(E_i - \mu)} + 1} . \tag{10.11}$$

Therefore, the expectation value of the particle number in this state is

$$\langle n_i \rangle = \frac{1}{\mathrm{e}^{\beta(E_i - \mu)} + 1} . \tag{10.12}$$

---

[2] For a system in which the particles interact, we cannot define single-particle states. The state of a particle is affected by all the other particles owing to the interaction, and therefore an individual particle cannot have a definite energy.

For a state with an energy $E_i \ll \mu - k_\mathrm{B}T$, we have $\langle n_i \rangle \simeq 1$, and for a state with an energy $E_i \gg \mu + k_\mathrm{B}T$, this expectation value is vanishingly small, i.e. $\langle n_i \rangle \simeq 0$. If we consider the right-hand side of the above equation as a function of E, we obtain the *Fermi distribution function* $f(E)$, where

$$f(E) = \frac{1}{\mathrm{e}^{\beta(E-\mu)} + 1} \,. \tag{10.13}$$

The behavior of $f(E)$ as a function of E is shown in Fig. 10.1.

The total number of particles in a macroscopic system is given almost exactly by

$$N = \sum_i \langle n_i \rangle \,. \tag{10.14}$$

This equation can be used to determine the value of μ when we consider a situation in which the total number of particles N is given. The grand partition function of the total system is

$$\Xi = \prod_i \Xi_i = \prod_i \left(1 + \mathrm{e}^{-\beta(E_i-\mu)}\right) \,. \tag{10.15}$$

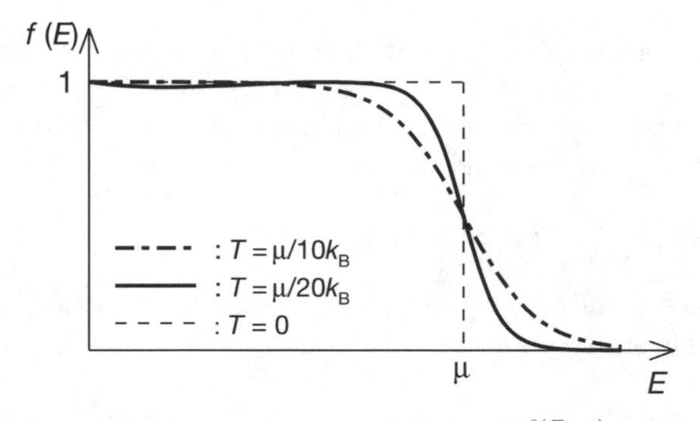

Fig. 10.1. The Fermi distribution function $f(E) = 1/[\mathrm{e}^{\beta(E-\mu)} + 1]$. The *thin dashed line* shows $f(E)$ at $T = 0$. The *solid* and *dash-dotted lines* show $f(E)$ at $k_\mathrm{B}T = \mu/20$ and $k_\mathrm{B}T = \mu/10$, respectively

Boson Gases

For a boson system, the ith single-particle state can accommodate any number of particles. The grand partition function for the ith state is

$$\begin{aligned} \Xi_i &= \mathrm{e}^0 + \mathrm{e}^{-\beta(E_i-\mu)} + \mathrm{e}^{-\beta(2E_i-2\mu)} + \mathrm{e}^{-\beta(3E_i-3\mu)} + \cdots \\ &= \frac{1}{1 - \mathrm{e}^{-\beta(E_i-\mu)}} \,. \end{aligned} \tag{10.16}$$

Here, we have used the fact that the energy of the ith state, when it contains n particles, is nE_i. The probability of having n particles in this state is

$$\frac{1}{\Xi}\mathrm{e}^{-\beta(nE_i-n\mu)} = \left[1-\mathrm{e}^{-\beta(E_i-\mu)}\right]\mathrm{e}^{-\beta(nE_i-n\mu)} . \qquad (10.17)$$

The expectation value of the number of particles in this state is

$$\begin{aligned}\langle n_i\rangle &= \sum_{n=1}^{\infty} n\mathrm{e}^{-n\beta(E_i-\mu)}\left[1-\mathrm{e}^{-\beta(E_i-\mu)}\right] \\ &= \frac{\mathrm{e}^{-\beta(E_i-\mu)}}{1-\mathrm{e}^{-\beta(E_i-\mu)}} \\ &= \frac{1}{\mathrm{e}^{\beta(E_i-\mu)}-1} . \end{aligned} \qquad (10.18)$$

If we consider the right-hand side of this equation as a function of E, we obtain the Bose distribution function

$$n(E) = \frac{1}{\mathrm{e}^{\beta(E-\mu)}-1} . \qquad (10.19)$$

We have already encountered this form, with $\mu = 0$, in various situations, namely the harmonic oscillations of diatomic molecules, the oscillations of a crystal lattice, and electromagnetic waves in a cavity. The behavior of the Bose distribution function is shown in Fig. 10.2. The grand partition function of the total system is

$$\Xi = \prod_i \Xi_i = \prod_i \frac{1}{1-\mathrm{e}^{-\beta(E_i-\mu)}} . \qquad (10.20)$$

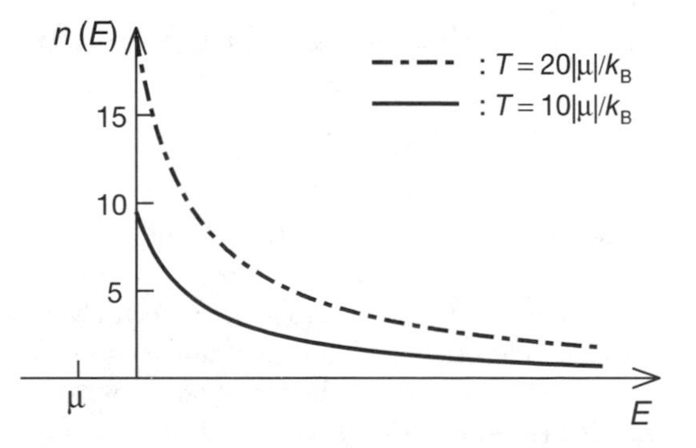

Fig. 10.2. The Bose distribution function $n(E) = 1/[\mathrm{e}^{\beta(E-\mu)}-1]$. The chemical potential μ must be negative; $n(E)$ at $k_\mathrm{B}T = 10|\mu|$ and $20|\mu|$ are shown by the *solid* and *dash-dotted lines*, respectively

10.3.2 Thermodynamic Functions

The free energy J has similar forms for the fermion gas and the boson gas:

$$\begin{aligned} J &= -k_{\mathrm{B}}T \ln \Xi \\ &= \mp k_{\mathrm{B}}T \sum_i \ln\left(1 \pm \mathrm{e}^{-\beta(E_i-\mu)}\right) , \end{aligned} \tag{10.21}$$

where the upper sign applies to fermions and the lower sign applies to bosons. The entropy can be obtained from this J:

$$\begin{aligned} S &= -\left(\frac{\partial J}{\partial T}\right)_{\mu} \\ &= \pm k_{\mathrm{B}} \sum_i \ln\left(1 \pm \mathrm{e}^{-\beta(E_i-\mu)}\right) + \frac{1}{T}\sum_i \frac{E_i - \mu}{\mathrm{e}^{\beta(E_i-\mu)} \pm 1} . \end{aligned} \tag{10.22}$$

Since the internal energy is given by

$$U = \sum_i \frac{E_i}{\mathrm{e}^{\beta(E_i-\mu)} \pm 1} , \tag{10.23}$$

this entropy can be rewritten as

$$S = -\frac{J}{T} + \frac{U}{T} - \frac{N\mu}{T} . \tag{10.24}$$

This is a relation expected from the general principle of thermodynamics $U = ST - PV + \mu N$. The entropy can also be written as

$$S = -k_{\mathrm{B}} \sum_i \left[\langle n_i\rangle \ln\langle n_i\rangle + (1 \mp \langle n_i\rangle) \ln\left(1 \mp \langle n_i\rangle\right)\right] . \tag{10.25}$$

10.4 Properties of a Free-Fermion Gas

In this section, we examine the properties of a Fermi gas using an explicit form for the single-particle energy E_i. It is known that liquid ^{3}He and the conduction electron systems in metals behave like a free-fermion gas. In fact, the interaction between helium atoms is not negligible and the electrons in metals interact quite strongly with each other through the Coulomb interaction, and so they themselves cannot behave like a free-fermion gas. However, each particle together with its influence it gives to the surroundings can be considered as a kind of a particle. This is a particle dressed with interaction effects, and usually called a quasiparticle. These quasiparticles can be considered as weakly interacting fermions [12].

We consider such quasiparticles and neglect the interaction between them. In a box of size $L_x \times L_y \times L_z$, a quasiparticle has the following energy:

$$E_i = \frac{\boldsymbol{p}_i^2}{2m} + g\mu_{\mathrm{B}} sB\,, \tag{10.26}$$

where m is the mass of a fermion, $\boldsymbol{p}_i$ is its momentum, g is the gyromagnetic ratio, also called the Landé g-factor,[3] $\mu_{\mathrm{B}} = e\hbar/2m$ is the Bohr magneton, $s = \pm 1/2$ is the spin of the fermion in units of $\hbar$, and B is the magnetic field applied to the system. The momentum is quantized such that it is specified by integers (n_x, n_y, n_z):

$$\boldsymbol{p}_i = \left(\frac{2\pi\hbar}{L_x} n_x, \frac{2\pi\hbar}{L_y} n_y, \frac{2\pi\hbar}{L_z} n_z \right). \tag{10.27}$$

Using the resulting energy spectrum, we define the density of single-particle states $D(E)$, such that $D(E)\,\mathrm{d}E$ is the number of single-particle states for each spin state in which the kinetic energy $E_{\mathrm{K}} = \boldsymbol{p}_i^2/2m$ is in the range E to $E + \mathrm{d}E$. Note that we have also defined a density of states in Chap. 2. The density of states in Chap. 2 is for an entire system, but that defined here is for single-particle states, and so instead of $\Omega(E)$ we write $D(E)$.[4]

Let us calculate the density of single-particle states. First we calculate the number of states for which the absolute value of the momentum p_i is in the range p to $p + \mathrm{d}p$. The volume of this range of p in momentum space is $4\pi p^2\,\mathrm{d}p$. On the other hand, the momentum is quantized, and the separation between adjacent momenta is $\Delta p_x = 2\pi\hbar/L_x$, and so on. Thus, there is one state in each volume $(2\pi\hbar)^3/L_xL_yL_z = (2\pi\hbar)^3/V$. Therefore, $4\pi p^2\,\mathrm{d}p$ divided by $(2\pi\hbar)^3/V$ gives the number of states in the momentum-space volume that we are considering, where V is the real-space volume of the system. That is, there are

$$\frac{4\pi p^2\,\mathrm{d}p}{(2\pi\hbar)^3/V} = \frac{V}{2\pi^2\hbar^3} p^2\,\mathrm{d}p \tag{10.28}$$

states. Using $\mathrm{d}E/\mathrm{d}p = p/2m$ and $p = \sqrt{2mE}$, we obtain

$$D(E)\,\mathrm{d}E = \frac{V}{2\pi^2\hbar^3} m\sqrt{2mE}\,\mathrm{d}E\,. \tag{10.29}$$

---

[3] As explained in Sect. 7.1, the magnetic moment of a particle is expected to be proportional to the angular momentum. However, the classical relation is not correct for elementary particles. The gyromagnetic ratio expresses the correction factor to be applied to the classical relation between these two quantities.

[4] Both the $D(E)$ defined here and $\Omega(E)$ are commonly called the density of states in the literature. It should not be difficult to determine from the context which density of states is meant.

Various thermodynamic variables can be written as a function of T, V, μ, and B in an integral form in terms of this single-particle density of states. We first note that in a magnetic field, a fermion with energy E_i has a kinetic energy $E_{\mathrm{K}} = E_i - g\mu_{\mathrm{B}}sB$. Therefore, the density of single-particle states at E_i is $D(E_i - g\mu_{\mathrm{B}}sB)$. We also note that the minimum value of E_i is $g\mu_{\mathrm{B}}sB$. Thus, the total number of particles N, the average energy in the magnetic field $\tilde{E}$, the total spin magnetic moment M, and the internal energy U are given by

$$\begin{aligned} N(T,V,\mu,B) &= \sum_i \sum_{s=\pm 1/2} \frac{1}{\mathrm{e}^{\beta(E_i-\mu)}+1} \\ &= \sum_{s=\pm 1/2} \int_{g\mu_{\mathrm{B}}sB}^{\infty} \mathrm{d}E\, D(E-g\mu_{\mathrm{B}}sB) \frac{1}{\mathrm{e}^{\beta(E-\mu)}+1}, \end{aligned} \tag{10.30}$$

$$\begin{aligned} \tilde{E}(T,V,\mu,B) &= \sum_i \sum_{s=\pm 1/2} \frac{E_i}{\mathrm{e}^{\beta(E_i-\mu)}+1} \\ &= \sum_{s=\pm 1/2} \int_{g\mu_{\mathrm{B}}sB}^{\infty} \mathrm{d}E\, D(E-g\mu_{\mathrm{B}}sB) \frac{E}{\mathrm{e}^{\beta(E-\mu)}+1}, \end{aligned} \tag{10.31}$$

$$\begin{aligned} M(T,V,\mu,B) &= \sum_i \sum_{s=\pm 1/2} \frac{-g\mu_{\mathrm{B}}s}{\mathrm{e}^{\beta(E_i-\mu)}+1} \\ &= \sum_{s=\pm 1/2} \int_{g\mu_{\mathrm{B}}sB}^{\infty} \mathrm{d}E\, D(E-g\mu_{\mathrm{B}}sB) \frac{-g\mu_{\mathrm{B}}s}{\mathrm{e}^{\beta(E-\mu)}+1}, \end{aligned} \tag{10.32}$$

and

$$\begin{aligned} U(T,V,\mu,M) &= \tilde{E}(T,V,\mu,B) + MB \\ &= \sum_{s=\pm 1/2} \int_{g\mu_{\mathrm{B}}sB}^{\infty} \mathrm{d}E\, D(E-g\mu_{\mathrm{B}}sB) \frac{E-g\mu_{\mathrm{B}}s}{\mathrm{e}^{\beta(E-\mu)}+1}. \end{aligned} \tag{10.33}$$

Finally, the pressure can be obtained from J, but it can also be written in terms of the internal energy U:

$$\begin{aligned} P(T,V,\mu,B)V &= -J = k_{\mathrm{B}}T \ln \Xi(T,V,\mu,B) \\ &= k_{\mathrm{B}}T \sum_i \sum_{s=\pm 1/2} \ln\left[1+\mathrm{e}^{-\beta(E_i-\mu)}\right] \\ &= k_{\mathrm{B}}T \sum_{s=\pm 1/2} \int_{g\mu_{\mathrm{B}}sB}^{\infty} \mathrm{d}E\, D(E-g\mu_{\mathrm{B}}sB) \ln\left[1+\mathrm{e}^{-\beta(E-\mu)}\right] \end{aligned}$$

$$
\begin{aligned}
&= k_{\mathrm{B}}T \frac{V}{2\pi^2\hbar^3} m\sqrt{2m} \sum_{s=\pm 1/2} \int_{g\mu_{\mathrm{B}}sB}^{\infty} \mathrm{d}E \sqrt{E - g\mu_{\mathrm{B}}sB} \\
&\quad \times \ln\left[1 + \mathrm{e}^{-\beta(E-\mu)}\right] \\
&= \frac{2}{3} \frac{V}{2\pi^2\hbar^3} m\sqrt{2m} \sum_{s=\pm 1/2} \int_{g\mu_{\mathrm{B}}sB}^{\infty} \mathrm{d}E \frac{[E - g\mu_{\mathrm{B}}sB]^{3/2}}{\mathrm{e}^{\beta(E-\mu)} + 1} \\
&= \frac{2}{3} \frac{V}{2\pi^2\hbar^3} m\sqrt{2m} \sum_{s=\pm 1/2} \int_{g\mu_{\mathrm{B}}sB}^{\infty} \mathrm{d}E\, D(E - g\mu_{\mathrm{B}}sB) \\
&\quad \times \frac{E - g\mu_{\mathrm{B}}sB}{\mathrm{e}^{\beta(E-\mu)} + 1} \\
&= \frac{2}{3} U . \qquad (10.34)
\end{aligned}
$$

This relationship between U and PV is generally satisfied for an ideal gas.

From (10.30), we can calculate $\mu(T, V, N, B)$, the chemical potential for a given value of the particle number N. Using this chemical potential, we can obtain the internal energy $U(T, V, N, M)$, the magnetization $M(T, V, N, B)$, and the pressure $P(T, V, N, B)$.

10.4.1 Properties at $T = 0$

$B = 0$

We can evaluate these variables analytically at $T = 0$. First we consider the case where there is no magnetic field, i.e. $B = 0$. In this case the spin-up ($s = 1/2$) and spin-down ($s = -1/2$) fermions have the same energy spectrum. The Fermi distribution function

$$
f(E) = \frac{1}{\mathrm{e}^{\beta(E-\mu)} + 1} \qquad (10.35)
$$

is unity for $0 \le E \le \mu$ and zero for $E > \mu$. Therefore, the total number of fermions N is given by

$$
\begin{aligned}
N &= 2 \int_0^{\mu} \mathrm{d}E\, D(E) = 2 \int_0^{\mu} \mathrm{d}E \frac{V}{2\pi^2\hbar^3} m\sqrt{2mE} \\
&= \frac{V}{3\pi^2\hbar^3} (2m\mu)^{3/2} . \qquad (10.36)
\end{aligned}
$$

At $T = 0$, μ has the meaning that it is the energy of the highest-energy occupied state when there are N fermions and the system is in the ground state. This highest energy is usually called the Fermi energy and written as E_F. That is, $E_F = \mu$ at $T = 0$. It can be expressed as follows:

$$E_F = \frac{\hbar^2}{2m}\left(\frac{3\pi^2 N}{V}\right)^{2/3} . \tag{10.37}$$

The absolute value of the momentum of a fermion with energy $E = E_F$ is called the Fermi momentum and written as p_F; it is equal to $\sqrt{2mE_F}$. The condition that $|\boldsymbol{p}| = p_F$ defines the surface of a sphere in momentum space. This sphere, which is shown in Fig. 10.3, is called the Fermi surface. Inside the Fermi surface, every single-particle state is occupied, and outside the surface, no state is occupied, at $T = 0$. The total number N can be expressed in terms of E_F and p_F as

$$N = \frac{V}{3\pi^2\hbar^3}(2mE_F)^{3/2} = \frac{V}{3\pi^2}\left(\frac{p_F}{\hbar}\right)^3 . \tag{10.38}$$

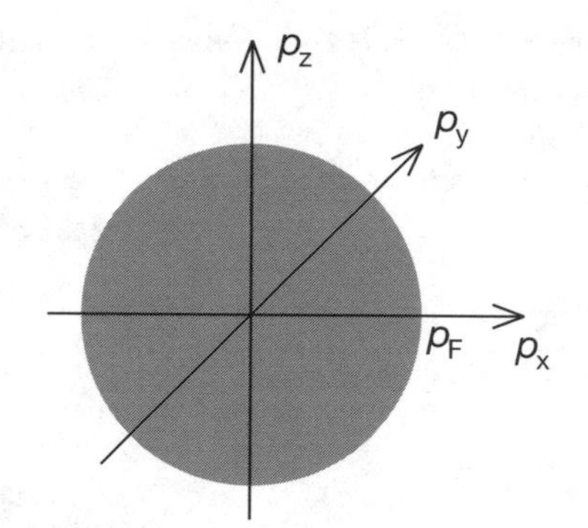

Fig. 10.3. The Fermi surface in momentum space. Every state with $|\boldsymbol{p}| \le p_F$ is occupied

We can now change the variable from μ to N, and consider a system containing a fixed number of fermions. In the calculation, it is convenient to use an expression for $D(E)$ in terms of the Fermi energy E_F,

$$D(E) = \frac{3}{4}N\left(\frac{E^{1/2}}{E_F^{3/2}}\right) . \tag{10.39}$$

The internal energy of the system is calculated as follows:

$$U = 2\int_0^{E_F} \mathrm{d}E\, E D(E) = \int_0^{E_F} \mathrm{d}E\, \frac{3}{2}N\left(\frac{E}{E_F}\right)^{3/2}$$
$$= \frac{3}{5}NE_F . \tag{10.40}$$

From the internal energy, we can calculate the pressure P at $T = 0$ and $B = 0$:

$$P = \frac{3U}{2V} = \frac{2N}{5V} E_{\rm F} \,. \tag{10.41}$$

Therefore, $k_{\rm B}T$ for the classical ideal gas is replaced by $2E_{\rm F}/5$, and the pressure is nonzero even at $T = 0$. These results for U and PV are consistent with (10.24):

$$\begin{aligned} ST &= -J + U - N\mu = \frac{2}{5}NE_{\rm F} + \frac{3}{5}NE_{\rm F} - NE_{\rm F} \\ &= 0 \,. \end{aligned} \tag{10.42}$$

To obtain the value of S at $T = 0$, we use (10.25): at $T = 0$, $\langle n_i \rangle = 0$ or 1, and so $\langle n_i \rangle \ln \langle n_i \rangle = 0$ and $(1 - \langle n_i \rangle) \ln(1 - \langle n_i \rangle) = 0$. Thus we conclude that $S = 0$.[5]

$B > 0$

Next we consider the effect of a magnetic field. Without a magnetic field, the up-spin states and down-spin states are equally occupied, and so the total magnetization of the system is zero. However, in the presence of a magnetic field, the single-particle energies for the same momentum $\boldsymbol{p}$ are split owing to the Zeeman term $g\mu_{\rm B}sB$. A magnetization is then induced by the magnetic field B. At $T = 0$, (10.30) reads

$$\begin{aligned} N(0, V, \mu, B) &= \sum_{s=\pm 1/2} \int_{g\mu_{\rm B}sB}^{\mu} \mathrm{d}E \, D(E - g\mu_{\rm B}sB) \\ &= \sum_{s=\pm 1/2} \int_{g\mu_{\rm B}sB}^{\mu} \mathrm{d}E \, \frac{3}{4} \frac{N}{E_{\rm F}^{3/2}} (E - g\mu_{\rm B}sB)^{1/2} \\ &= \sum_{s=\pm 1/2} \frac{1}{2} \frac{N}{E_{\rm F}^{3/2}} (\mu - g\mu_{\rm B}sB)^{3/2} \\ &= N \left(\frac{\mu}{E_{\rm F}} \right)^{3/2} + O(B^2) \,. \end{aligned} \tag{10.43}$$

Thus, $\mu = E_{\rm F} + O(B^2)$. For the calculation of the magnetic moment, we need only terms linear in B, and so we can neglect the dependence of the chemical potential μ on B.

Now we calculate the total magnetic moment. From (10.32), we obtain

$$M = \sum_{s=\pm 1/2} \int_{g\mu_{\rm B}sB}^{E_{\rm F}} \mathrm{d}E \, (-g\mu_{\rm B}s) D(E - g\mu_{\rm B}sB) \,. \tag{10.44}$$

[5] The fact that liquid ^{3}He has a very small entropy at low T, even though each atom has a magnetic moment, has been remarked on in Sect. 8.4.

In order to obtain the magnetic susceptibility, we evaluate this integral up to only the first order in B. We obtain

$$M \simeq \frac{1}{2}\left(g\mu_{\mathrm{B}}\right)^2 D(E_{\mathrm{F}})B = \frac{3}{2}\left(\frac{1}{2}g\mu_{\mathrm{B}}\right)^2 \frac{N}{E_{\mathrm{F}}}B\,. \tag{10.45}$$

Thus the magnetic susceptibility per unit volume is

$$\chi = \frac{3}{2}\left(\frac{1}{2}g\mu_{\mathrm{B}}\right)^2 \frac{\mu_0 N}{V E_{\mathrm{F}}}\,, \tag{10.46}$$

where μ_0 is the permeability of free space.

10.4.2 Properties at Low Temperature

The Fermi Temperature

At nonzero temperatures, thermodynamic variables such as the internal energy and the magnetic susceptibility deviate from their values at $T = 0$. The amount of deviation depends on whether we consider the system at a fixed number of particles or a fixed value of the chemical potential. Here, we calculate the deviation at a fixed number of particles in the lowest order of the temperature T. Since the Fermi distribution function is expressed in terms of the chemical potential, we first calculate how the chemical potential changes at nonzero temperature, and then calculate the temperature dependence of the internal energy, the heat capacity, and the magnetic susceptibility of the system at a fixed number of particles.

In order to present the results in a transparent form, we need a natural scale for the temperature. This scale is provided by the Fermi temperature T_{F}, which is defined as the Fermi energy divided by the Boltzmann constant k_{B}:

$$T_{\mathrm{F}} = \frac{E_{\mathrm{F}}}{k_{\mathrm{B}}} = \frac{\hbar^2}{2mk_{\mathrm{B}}}\left(\frac{3\pi^2 N}{V}\right)^{2/3}. \tag{10.47}$$

The Fermi energy or the Fermi temperature is a function of the particle density N/V. If the conduction electrons in a metal are considered as free fermions, we can estimate the Fermi temperature of the metal. From the mass density of the metal and the mass number of the constituent atoms, we can calculate the number density of atoms. From the valence of the metal, we know how many conduction electrons are provided to the metal by each atom. We can then estimate the conduction electron density N/V. The effective mass m is usually of the order of the vacuum mass of an electron. So, by assuming that the effective mass is equal to the vacuum mass, we can estimate the Fermi energy and the Fermi temperature. For most metals, this procedure gives values of T_{F} between 10^4 and 10^5 K. Some values of the Fermi temperature for typical metals are given in Table 10.1. In this subsection, "low temperature" means that $T \ll T_{\mathrm{F}}$. Therefore, metals at room temperature can be regarded as being in the low-temperature regime.

Table 10.1. Fermi temperatures T_{F} of typical metals

Metal	Valence	Fermi temperature (10^4 K)
Li	1	5.48
Na	1	3.75
K	1	2.46
Cu	1	8.12
Ag	1	6.36
Au	1	6.39
Mg	2	8.27
Al	3	13.49
Pb	4	10.87

Formulas for the Temperature Dependence of Thermodynamic Variables

The calculation of the temperature dependence of the thermodynamic variables for a system with a fixed number of particles is done in two steps. First, the temperature dependence for a system at a given chemical potential μ is calculated. One of the thermodynamic variables is the particle number N. Then $N(T,\mu)$, considered as a function of T and μ, is inverted to obtain $\mu(T,N)$. In a second step, this $\mu(T,N)$ is put into the expressions for the other variables to obtain the temperature dependence for fixed N.

The first step is performed by use of a general formula derived below. For simplicity, we first consider a system in the absence of a magnetic field. In this case thermodynamic variables such as N and U are given by integrals of the following kind:

$$F(T,V,\mu) = \int_{\epsilon_0}^{\infty} \mathrm{d}E\, g(E) f(E)\,, \tag{10.48}$$

where F is one such variable, $\epsilon_0 = 0$ is the lowest single-particle energy,[6] $g(E)$ depends on which variable we want to calculate, and $f(E)$ is the Fermi distribution function. This type of integral can be evaluated at low temperature as a power series in the temperature T. In order to obtain the desired result, we first define a function $G(E)$, which is the integral of $g(E)$:

$$G(E) \equiv \int_{\epsilon_0}^{E} g(E')\,\mathrm{d}E'\,. \tag{10.49}$$

The value of $F(T,V,\mu)$ at $T=0$ is the value of G at $E=\mu$:

$$F(0,V,\mu) = G(\mu)\,. \tag{10.50}$$

[6] We write the lowest energy as ϵ_0 in these equations. At $B=0$, $\epsilon_0=0$. However, at finite magnetic field, this energy becomes finite, because $\epsilon_0 = g\mu_{\mathrm{B}} sB$.

Therefore, $G(\mu)$ is the zeroth-order term in the expansion of $F(T,V,\mu)$ as a power series in T. Using this $G(E)$, we can rewrite the integral for $F(T,V,\mu)$:

$$\begin{aligned} F(T,V,\mu) &= \int_{\epsilon_0}^{\infty} \mathrm{d}E \left[\frac{\mathrm{d}}{\mathrm{d}E} G(E)\right] f(E) \\ &= \Big[G(E)f(E)\Big]_{\epsilon_0}^{\infty} - \int_{\epsilon_0}^{\infty} \mathrm{d}E\, G(E) \frac{\mathrm{d}f(E)}{\mathrm{d}E} . \end{aligned} \tag{10.51}$$

The first term in the last line is zero, because at $E = \epsilon_0$, $G(\epsilon_0) = 0$ and $f(\epsilon_0) \simeq 1$, and as $E \to \infty$, $f(E) \simeq \mathrm{e}^{-\beta(E-\mu)}$ vanishes exponentially. To evaluate the second term, we notice that $\mathrm{d}f(E)/\mathrm{d}E$ is sharply peaked around $E = \mu$ and vanishes rapidly as E deviates from μ, as shown in Fig. 10.4. This fact enables us to do the following:

1. Extend the lower bound of the integral to $-\infty$.
2. Expand the function $G(E)$ around μ and retain only the low-order terms.

Since

$$G(E) \simeq G(\mu) + (E-\mu)G'(\mu) + \frac{1}{2}(E-\mu)^2 G''(\mu) + \cdots , \tag{10.52}$$

we need to evaluate the following integral for $n = 0, 1, 2, \cdots$:

$$\int_{-\infty}^{\infty} \mathrm{d}E\, (E-\mu)^n \frac{\mathrm{d}f(E)}{\mathrm{d}E} = \beta^{-n} \int_{-\infty}^{\infty} \mathrm{d}x\, x^n \frac{\mathrm{d}}{\mathrm{d}x}\left(\frac{1}{\mathrm{e}^x+1}\right) . \tag{10.53}$$

For $n = 0$, this integral is easily evaluated, and the result is -1. For odd n, the integral vanishes, since

$$\frac{\mathrm{d}}{\mathrm{d}x}\left(\frac{1}{\mathrm{e}^x+1}\right) = -\frac{1}{4}\mathrm{sech}^2\frac{x}{2} \tag{10.54}$$

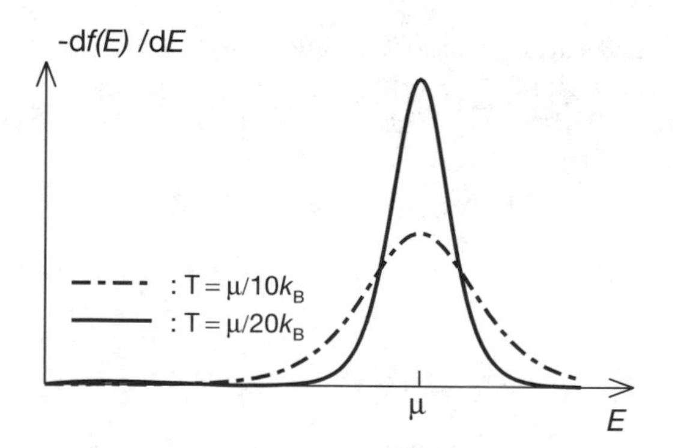

Fig. 10.4. The derivative of the Fermi distribution function, $\mathrm{d}f(E)/\mathrm{d}E$; $-\mathrm{d}f(E)/\mathrm{d}E$ is shown for $k_\mathrm{B}T = \mu/20$ and $\mu/10$ by the *solid* and *dash-dotted lines*, respectively. This derivative is sharply peaked around $E = \mu$ at low temperature

is an even function of x. For even n, it can be calculated as follows:

$$\begin{aligned}\beta^{-n}\int_{-\infty}^{\infty}\mathrm{d}x\,x^n\frac{\mathrm{d}}{\mathrm{d}x}\left(\frac{1}{\mathrm{e}^x+1}\right) &= 2\beta^{-n}\int_0^{\infty}\mathrm{d}x\,x^n\frac{\mathrm{d}}{\mathrm{d}x}\left(\frac{1}{\mathrm{e}^x+1}\right)\\ &= -2n\beta^{-n}\int_0^{\infty}\mathrm{d}x\,\frac{x^{n-1}}{\mathrm{e}^x+1}\\ &= -2n\beta^{-n}\sum_{k=0}^{\infty}(-1)^k\int_0^{\infty}\mathrm{d}x\,x^{n-1}\mathrm{e}^{-(k+1)x}\\ &= -2n!\beta^{-n}\sum_{k=0}^{\infty}\frac{(-1)^k}{(k+1)^n}\\ &= -2n!\beta^{-n}(1-2^{1-n})\zeta(n)\,. \end{aligned} \tag{10.55}$$

Here $\zeta(n)=\sum_{k=1}^{\infty}k^{-n}$ is the Riemann zeta function, which has the following values:[7]

$$\zeta(2)=\frac{\pi^2}{6},\quad \zeta(4)=\frac{\pi^4}{90}\,. \tag{10.56}$$

From these equations, (10.48) can be expanded as a series in T:

$$F(T)=G(\mu)+\frac{(\pi k_{\mathrm{B}}T)^2}{6}g'(\mu)+\frac{7(\pi k_{\mathrm{B}}T)^4}{360}g'''(\mu)+\cdots\,. \tag{10.57}$$

We shall calculate the heat capacity and the magnetic susceptibility as functions of the number of fermions N and of T at low temperature using this formula.

Chemical Potential at $B = 0$

Now we shall calculate the particle number $N(T,V,\mu)$ to obtain $\mu(T,V,N)$ in the absence of a magnetic field. The magnetic-field dependence is considered later, when we calculate the susceptibility. From (10.30), we obtain

$$N(T,V,\mu)=2\int_0^{\infty}\mathrm{d}E\,D(E)f(E)\,. \tag{10.58}$$

Thus, in this case

$$g(E)=2D(E)=\frac{3}{2}\frac{N}{E_{\mathrm{F}}}\sqrt{\frac{E}{E_{\mathrm{F}}}}\,, \tag{10.59}$$

$$G(\mu)=N\left(\frac{\mu}{E_{\mathrm{F}}}\right)^{3/2}\,, \tag{10.60}$$

[7] The Riemann zeta function is described in Appendix G.

and

$$g'(\mu) = \frac{3}{4} \frac{N}{E_{\mathrm{F}}^{3/2}} \frac{1}{\mu^{1/2}} . \tag{10.61}$$

Using these equations, we obtain $N(T, V, \mu)$ up to the second order in T:

$$N(T, V, \mu) \simeq N \left(\frac{\mu}{E_{\mathrm{F}}} \right)^{3/2} + \frac{1}{8} \frac{N(\pi k_{\mathrm{B}} T)^2}{E_{\mathrm{F}}^{3/2} \mu^{1/2}} . \tag{10.62}$$

As we expect that $\mu/E_{\mathrm{F}} = 1 + O(T^2)$, μ in the second term on the right-hand side can be replaced by E_{F}. Then μ/E_{F} can be obtained as follows up to the second order in T/T_{F}:

$$\frac{\mu}{E_{\mathrm{F}}} \simeq \left[1 - \frac{\pi^2}{8} \left(\frac{T}{T_{\mathrm{F}}} \right)^2 \right]^{2/3} \simeq 1 - \frac{\pi^2}{12} \left(\frac{T}{T_{\mathrm{F}}} \right)^2 . \tag{10.63}$$

In this equation, we have used the relation that $E_{\mathrm{F}} = k_{\mathrm{B}} T_{\mathrm{F}}$. This temperature dependence and the correct temperature dependence of μ up to $T = 3T_{\mathrm{F}}$ are shown in Fig. 10.5.

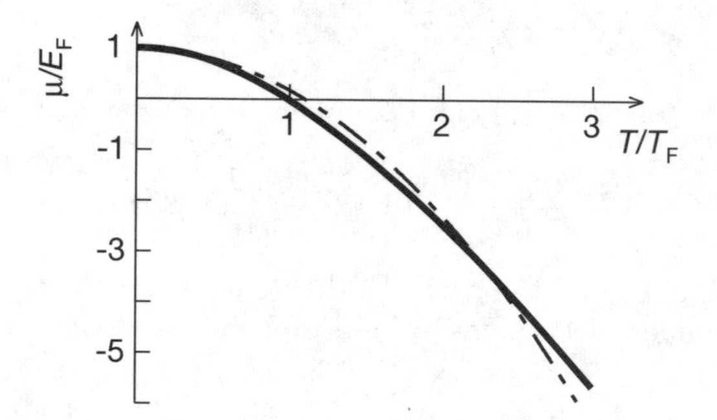

Fig. 10.5. Temperature dependence of μ/E_{F} as a function of T/T_{F}. The exact result is shown by the *solid line*, and the *dash-dotted line* shows the approximation to the chemical potential (10.63)

Internal Energy, Pressure, and Heat Capacity

The internal energy in the absence of a magnetic field is

$$U(T, V, \mu) = \int_0^\infty \mathrm{d}E \, 2ED(E) f(E) . \tag{10.64}$$

Thus, in this case

$$g(E) = 2ED(E) = \frac{3}{2} N \left(\frac{E}{E_{\mathrm{F}}} \right)^{3/2} , \tag{10.65}$$

$$G(\mu) = \frac{3}{5} N \frac{\mu^{5/2}}{E_{\mathrm{F}}^{3/2}} , \tag{10.66}$$

and

$$g'(\mu) = \frac{9}{4} N \frac{\mu^{1/2}}{E_F^{3/2}}. \tag{10.67}$$

Using these relations, we obtain $U(T, V, \mu)$:

$$U(T, V, \mu) = \frac{3}{5} N \frac{\mu^{5/2}}{E_F^{3/2}} + \frac{3\pi^2}{8} N \frac{(k_B T)^2 \mu^{1/2}}{E_F^{3/2}}. \tag{10.68}$$

The internal energy at a given value of N, $U(T, V, N)$, is obtained by using the expression for the chemical potential (10.63):

$$\begin{aligned} U(T, V, N) &\simeq \frac{3}{5} N E_F \left[1 - \frac{\pi^2}{12} \left(\frac{T}{T_F} \right)^2 \right]^{5/2} + \frac{3\pi^2}{8} N E_F \left(\frac{T}{T_F} \right)^2 \\ &\simeq \frac{3}{5} N E_F \left[1 - \frac{5\pi^2}{24} \left(\frac{T}{T_F} \right)^2 \right] + \frac{3\pi^2}{8} N E_F \left(\frac{T}{T_F} \right)^2 \\ &= \frac{3}{5} N E_F \left[1 + \frac{5\pi^2}{12} \left(\frac{T}{T_F} \right)^2 \right]. \end{aligned} \tag{10.69}$$

The pressure of the system is related to U by $PV = (2/3)U$:

$$P(T, V, N)V \simeq \frac{2}{5} N E_F \left[1 + \frac{5\pi^2}{12} \left(\frac{T}{T_F} \right)^2 \right]. \tag{10.70}$$

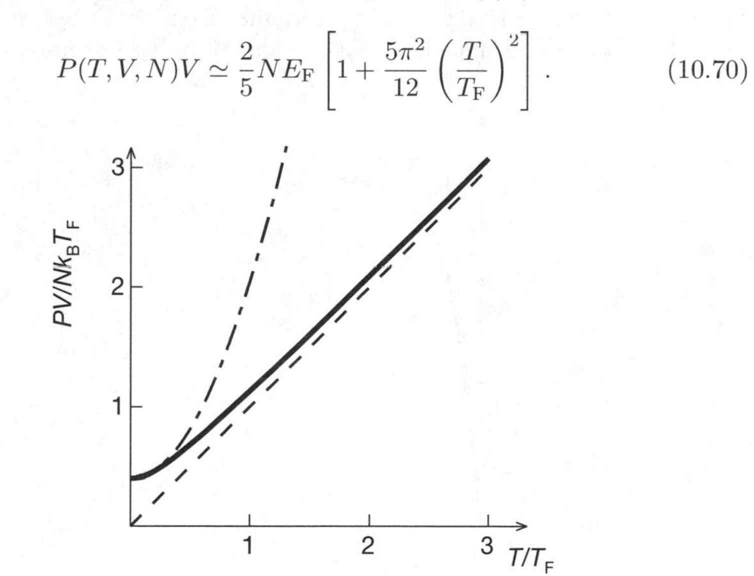

Fig. 10.6. Temperature dependence of PV/Nk_BT_F as a function of T/T_F. The *solid line* shows the correct temperature dependence, the *dash-dotted line* shows the approximation (10.70), and the *dashed line* shows $PV = Nk_BT$, which is the relation for a classical ideal gas

The temperature dependence of PV is shown in Fig. 10.6, where the solid line shows the correct temperature dependence, the dash-dotted line shows the approximation (10.70), and the dashed line shows the classical result $PV = Nk_{\rm B}T$.

The fact that the correction to the internal energy is proportional to T^2 is easily understood. The Fermi distribution function deviates from the distribution function at $T = 0$ only near the Fermi energy. Only particles with energies near the Fermi energy can have a thermal energy $k_{\rm B}T$, and the number of such particles is of the order of $k_{\rm B}T$ times the density of states $D(E_{\rm F})$. Therefore, the increase in the internal energy at finite temperature is proportional to T^2.

Now that we have obtained the internal energy, we can calculate the heat capacity of a free-fermion gas at low temperature:

$$C(T,V,N) = \left(\frac{\partial U(T,V,N)}{\partial T}\right)_{V,N} \simeq \frac{\pi^2}{2} N k_{\rm B} \frac{T}{T_{\rm F}} . \tag{10.71}$$

Since $C = T(\partial S/\partial T)$, and $S(0,V,N) = 0$ at $T = 0$, we can also obtain the entropy at low temperature:

$$S(T,V,N) \simeq C(T,V,N) \simeq \frac{\pi^2}{2} N k_{\rm B} \frac{T}{T_{\rm F}} . \tag{10.72}$$

We can compare this result with that obtained from the Debye model of lattice vibrations. The latter model predicts a value of $3Nk_{\rm B}$ at temperatures higher

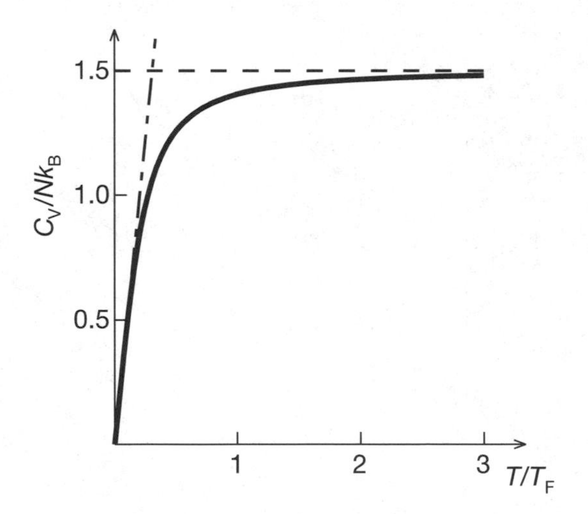

Fig. 10.7. Temperature dependence of $C_V/Nk_{\rm B}$ as a function of $T/T_{\rm F}$. The *solid line* shows the correct temperature dependence, the *dash-dotted line* shows the lowest-order approximation (10.71), and the *dashed line* shows the classical result, $C_{\rm V} = 1.5Nk_{\rm B}$

than the Debye temperature. For ordinary solids, this high-temperature condition is satisfied at room temperature. On the other hand, a conduction electron system has a T_{F} of the order of 10^4 K, and room temperature ($\simeq 300\,\mathrm{K}$) is much lower than T_{F}. Therefore, the contribution to the heat capacity of a metal from the conduction electrons is negligibly small at room temperature. We note also that at temperatures much higher than T_{F}, a fermion gas behaves as a classical ideal gas. Therefore, at such temperatures the constant-volume heat capacity tends to $(3/2)Nk_{\mathrm{B}}$. The temperature dependence of the heat capacity is shown in Fig. 10.7.

Magnetic Susceptibility

The magnetic susceptibility due to the spins is obtained from

$$\chi(T,V,N) = \lim_{B\to 0} \frac{\mu_0 M(T,V,N,B)}{BV} \,.$$

Thus we need to calculate the total magnetic moment $M(T,V,N,B)$ and $\mu(T,V,N,B)$ up to first order in the magnetic field B. First, we need to examine how the chemical potential depends on the magnetic field. For that purpose, we write down an equation for the number of particles at given μ in the presence of a magnetic field:

$$N(T,V,\mu,B) = \sum_{s=\pm 1/2} \int_{g\mu_{\mathrm{B}}sB}^{\infty} \mathrm{d}E\, D(E - g\mu_{\mathrm{B}}sB) f(E) \,. \tag{10.73}$$

The right hand side of this equation is clearly a continuous even function of B. Therefore, without actual calculation, we can guess the following result:

$$\begin{aligned} N(T,V,\mu,B) &= N(T,V,\mu,0) + O\left(B^2\right) \\ &\simeq N\left(\frac{\mu}{E_{\mathrm{F}}}\right)^{3/2} + \frac{1}{8} N \frac{(\pi k_{\mathrm{B}} T)^2}{E_{\mathrm{F}}^{3/2} \mu^{1/2}} + O\left(B^2\right) \,. \end{aligned} \tag{10.74}$$

The correction to the chemical potential then also starts from the second order in B, namely

$$\begin{aligned} \mu(T,V,N,B) &= \mu(T,V,N,0) + O\left(B^2\right) \\ &\simeq E_{\mathrm{F}}\left[1 - \frac{\pi^2}{12}\left(\frac{T}{T_{\mathrm{F}}}\right)^2\right] + O\left(B^2\right) \,. \end{aligned} \tag{10.75}$$

Therefore, for the calculation of the magnetic susceptibility, we need not consider the magnetic-field dependence of the chemical potential.

Next we calculate the total spin magnetic moment to the lowest order in B. First, we calculate it at a given chemical potential μ. It is given by the following equation:

$$
\begin{aligned}
M(T,V,\mu,B) &= \sum_{s=\pm 1/2} \int_{g\mu_{\mathrm{B}}sB}^{\infty} \mathrm{d}E\,(-g\mu_{\mathrm{B}}s)D\,(E-g\mu_{\mathrm{B}}sB)\,f(E) \\
&\simeq -g\mu_{\mathrm{B}} \sum_{s=\pm 1/2} \left[\int_0^{\mu-g\mu_{\mathrm{B}}sB} \mathrm{d}E\, sD(E) \right. \\
&\quad \left. + \frac{\pi^2}{6}\left(k_{\mathrm{B}}T\right)^2 sD'\left(\mu - g\mu_{\mathrm{B}}sB\right) + O\left(T^4\right) \right] \\
&\simeq \frac{3}{2}\left(\frac{1}{2}g\mu_{\mathrm{B}}\right)^2 \frac{BN}{E_{\mathrm{F}}} \left[\left(\frac{\mu}{E_{\mathrm{F}}}\right)^{1/2} - \frac{\pi^2}{24}\left(\frac{T}{T_{\mathrm{F}}}\right)^2 \left(\frac{E_{\mathrm{F}}}{\mu}\right)^{3/2} \right] .
\end{aligned}
\tag{10.76}
$$

Inserting the expression for the chemical potential, we obtain the total magnetic moment up to the term linear in B and up to terms in T^2:

$$
M(T,V,N,B) = \frac{3}{2}\left(\frac{1}{2}g\mu_{\mathrm{B}}\right)^2 \frac{BN}{E_{\mathrm{F}}} \left[1 - \frac{\pi^2}{12}\left(\frac{T}{T_{\mathrm{F}}}\right)^2 \right] . \tag{10.77}
$$

Finally, we obtain the magnetic susceptibility per unit volume:

$$
\begin{aligned}
\chi(T,V,N) &= \lim_{B\to 0}\left(\frac{\mu_0 M}{BV}\right) \\
&= \frac{3}{2}\left(\frac{1}{2}g\mu_{\mathrm{B}}\right)^2 \frac{\mu_0 N}{E_{\mathrm{F}}V} \left[1 - \frac{\pi^2}{12}\left(\frac{T}{T_{\mathrm{F}}}\right)^2 \right] .
\end{aligned}
\tag{10.78}
$$

The susceptibility decreases quadratically as $(T/T_{\mathrm{F}})^2$ from its value at $T=0$. Again, the deviation from the zero-temperature value is small for ordinary metals at room temperature. At temperatures much higher than T_{F}, χ tends to the classical value for N independent particles in a volume V, (7.18):

$$
\chi = \left(\frac{1}{2}g\mu_{\mathrm{B}}\right)^2 \left(\frac{\mu_0 N}{k_{\mathrm{B}}TV}\right) . \tag{10.79}
$$

The reason that χ saturates at a finite value as $T \to 0$ can be understood from the behavior of the Fermi distribution function. Namely, only particles that have energies around the Fermi energy are active at low temperature and can change their spin state in response to an external magnetic field. The number of particles can be estimated as $2D(E_{\rm F})k_{\rm B}T$. We should replace N in (10.79) with this estimate. The correct value at $T = 0$,

$$\chi = \frac{3}{2}\left(\frac{1}{2}g\mu_{\rm B}\right)^2 \frac{\mu_0 N}{E_{\rm F}V},$$

is then obtained.

10.5 Properties of a Free-Boson Gas

10.5.1 The Two Kinds of Bose Gas

As we have seen in Chap. 5, the Bose distribution function $n(E)$ with $\mu = 0$ appears when we consider either the Debye model of a solid or black-body radiation. In these cases, a wave with a wave vector $\boldsymbol{k}$ and angular frequency $\omega(\boldsymbol{k}) = c|\boldsymbol{k}|$ is allowed to have quantized energies

$$E(\boldsymbol{k}) = \left(n + \frac{1}{2}\right)\hbar\omega(\boldsymbol{k}),$$

and the expectation value of n is given by the Bose distribution function with $\mu = 0$. This result can be interpreted as showing that lattice vibrations and electromagnetic radiation behave as Bose particles, which have a momentum $\boldsymbol{p} = \hbar\boldsymbol{k}$ and an energy $E = \hbar\omega(\boldsymbol{k})$. The particles for lattice vibrations are called phonons, and those for electromagnetic radiation are called photons.

The total number of these bosons is not fixed. They are created by the energy of the heat bath, and absorbed as energy into the heat bath. This nonconservation of the total number of particles is the origin of the condition that $\mu = 0$ for these bosons. The chemical potential μ is defined as the derivative of the entropy of the heat bath with respect to the number of particles in the heat bath. Once phonons or photons have been absorbed into the heat bath, they lose their identity, and so the entropy of the heat bath does not depend on the number of particles absorbed. Thus the chemical potential must be zero.

We have already studied the properties of a gas made up of these nonconserved particles in Chap. 5. What we consider in this section is a second kind of Bose gas, made up of particles for which the total number of particles is conserved. An ordinary gas for which the constituent atoms or molecules are bosons is a natural example of such a Bose system.

The characteristic properties of a Bose gas appear at low temperature. So we need a system that behaves like a free-boson gas even at low temperature. One candidate is helium-4. The helium-4 atom is a Bose particle, and helium gas is hard to liquefy. At 1 atm, it liquefies at 4.2 K. This is the lowest gas–liquid phase transition temperature of any substance except for helium-3. At atmospheric pressure, all other substances solidify at much higher temperatures. However, 4.2 K is still not low enough for a Bose gas to show its most peculiar characteristics. On the other hand, liquid helium-4 remains a liquid down to $T \to 0$ below about 25 atm. Even though it is a liquid and the interactions between the atoms are not negligible, some aspects of the liquid, such as the superfluid phase transition, can be understood as a property of a Bose gas. Other, much better examples of Bose gases have recently been realized: these are provided by vapors of alkali metals, cooled down to temperatures in the microkelvin range. We are now going to study the properties of such a Bose gas at low temperature.

Since the Bose distribution function diverges if the chemical potential coincides with a single-particle energy of the Bose gas, μ must be lower than the lowest single-particle energy, which is zero for gas of atoms. Therefore, the chemical potential must always be negative at finite temperature for the second kind of Bose gas.

10.5.2 Properties at Low Temperature

Bose–Einstein Condensation

The energy spectrum of an atomic Bose particle is the same as that for a Fermi gas, and so the density of single-particle states has the same form:

$$D(E)\,\mathrm{d}E = \frac{V}{2\pi^2\hbar^3} m\sqrt{2mE}\,\mathrm{d}E\,.$$

A Bose particle has an integer spin. Here we consider the simplest case, namely that of particles with spin zero. In this case the number of particles is given by the following equation:

$$N = \sum_i \frac{1}{e^{\beta(E_i-\mu)} - 1}\,. \tag{10.80}$$

At high temperature, we can calculate this summation as an integral with respect to the energy:

$$N = \int_0^\infty \mathrm{d}E\, D(E) \frac{1}{e^{\beta(E-\mu)} - 1}\,. \tag{10.81}$$

However, at low temperature, this replacement is not permissible. This is a special property of a Bose gas. We shall see in the following why this replacement is not correct.

The Fermi distribution function is finite for any energy, and is less than or equal to one. On the other hand, the Bose distribution function can diverge as

$\mu \to E_i$, whereas it becomes vanishingly small when $E_i - \mu \gg k_B T$. Therefore, at $T = 0$, almost all the states satisfy $E_i - \mu \gg k_B T = 0$, and all the particles are accommodated in the lowest-energy single-particle state at $E_i = 0$. The chemical potential is vanishingly small, and the internal energy is zero. That is, $\mu \to 0$ and $U = 0$ at $T = 0$.

When the temperature is low enough, we can expect that the majority of the bosons will still occupy the single-particle state at $E = 0$, and only some of the bosons will be excited to finite-energy states. This phase, where a finite fraction of the bosons occupies the lowest-energy single-particle state, is called a Bose–Einstein condensed state, and this phenomenon is called Bose–Einstein condensation. The bosons in the lowest-energy single-particle state are called the Bose–Einstein condensate.

In a Bose–Einstein condensed state, it is not permissible to replace the summation in (10.80) by the integral form (10.81). This is because the density of states $D(E)$ vanishes at $E = 0$, in spite of the fact that the lowest-energy single-particle state exists at $E = 0$. In (10.81), the contribution from the state at $E = 0$ has not been counted, even though most bosons are accommodated in this state. However, we need to replace the summation by an integral to perform an analytical investigation of boson systems at low temperature.

This is accomplished by treating the state at $E = 0$ separately, and treating the contribution from the other states by means of an integral. That is, we calculate the summation as follows in the case of a Bose–Einstein condensed phase:

$$N = \sum_i \frac{1}{e^{\beta E_i} - 1}$$
$$= N_0 + \int_0^\infty dE\, D(E) \frac{1}{e^{\beta E} - 1} . \tag{10.82}$$

In this equation, N_0 is the number of bosons in the $E = 0$ state, or the number in the Bose–Einstein condensate. We have put $\mu = 0$ in this equation, because it has only a small negative value in the condensed phase. The actual value is given below. A very small value of μ is necessary to give us $N_0 = O(N)$ at $E = 0$.

For a better understanding of Bose–Einstein condensation, let us consider the system at higher temperature, and then see what happens as the temperature is lowered. At high temperature, the bosons are distributed among various single-particle states, and no state contains a macroscopic number of bosons. Thus, for any state, E_i is higher than μ, and hence μ must have a finite negative value, which is determined by (10.81) for a given particle number N. Now, what happens when the temperature is lowered? The right-hand sides of (10.80) and (10.81) are monotonically increasing functions of μ, and also increasing functions of T. Thus, as the temperature is lowered at fixed N, the chemical potential increases towards zero. If we use the original summation in (10.80) to determine the chemical potential, we can always find a solution: the right-hand side tends to zero as $\mu \to -\infty$, and it diverges as $\mu \to 0$. However, it turns out that the right-hand side of the integral in (10.81) remains finite even at $\mu = 0$, and the limiting value decreases as the temperature is lowered. The temperature at which the limiting value coincides with N is the critical temperature T_c for Bose–Einstein condensation. For $T > T_\mathrm{c}$, the right-hand side of (10.81) at $\mu = 0$ is larger than the number of bosons, and the finite negative chemical potential can be determined. On the other hand, for $T < T_\mathrm{c}$, the right-hand side is less than N and we must put $\mu = 0$ and use (10.82) with a finite N_0.

Now let us determine the critical temperature T_c. From the considerations above, we know that it is determined by the following equation:

$$N = \int_0^\infty \mathrm{d}E\, D(E) \frac{1}{\mathrm{e}^{\beta_\mathrm{c} E} - 1}, \tag{10.83}$$

where $\beta_\mathrm{c} \equiv 1/k_\mathrm{B} T_\mathrm{c}$. The right-hand side is

$$\begin{aligned}
\text{r.h.s} &= \frac{V}{4\pi^2\hbar^3}(2m)^{3/2} \int_0^\infty \mathrm{d}E\, \frac{\sqrt{E}}{\mathrm{e}^{\beta_\mathrm{c} E} - 1} \\
&= \frac{V}{4\pi^2\hbar^3}\left(\frac{2m}{\beta_\mathrm{c}}\right)^{3/2} \int_0^\infty \mathrm{d}x\, \frac{x^{1/2}}{\mathrm{e}^{x} - 1} \\
&= \frac{V}{4\pi^2\hbar^3}\left(\frac{2m}{\beta_\mathrm{c}}\right)^{3/2} \Gamma\left(\frac{3}{2}\right)\zeta\left(\frac{3}{2}\right) \\
&= V\left(\frac{m k_\mathrm{B} T_\mathrm{c}}{2\pi\hbar^2}\right)^{3/2} \zeta\left(\frac{3}{2}\right),
\end{aligned} \tag{10.84}$$

where the integral representation of the zeta function (G.2) has been used. Using the value of the zeta function $\zeta(3/2) = 2.612\ldots$, we can determine the critical temperature:

$$T_\mathrm{c} = \frac{2\pi\hbar^2}{m k_\mathrm{B}}\left(\frac{N}{2.612V}\right)^{2/3}. \tag{10.85}$$

Below T_c, the number of particles in the Bose–Einstein condensate becomes of the order of N:

$$N_0 = N - V\left(\frac{m k_\mathrm{B} T}{2\pi\hbar^2}\right)^{3/2} \zeta\left(\frac{3}{2}\right) = N\left[1 - \left(\frac{T}{T_\mathrm{c}}\right)^{3/2}\right]. \tag{10.86}$$

From this value of N_0, we can determine the value of the chemical potential at $T < T_\mathrm{c}$. It satisfies the condition that

$$N_0 = \frac{1}{\mathrm{e}^{-\beta\mu} - 1} \simeq -\frac{k_\mathrm{B}T}{\mu} . \tag{10.87}$$

Here we have used the fact that $|\beta\mu| \ll 1$ and $\mathrm{e}^{-\beta\mu} \simeq 1 - \beta\mu$. Thus,

$$\mu = -\frac{k_\mathrm{B}T}{N\left[1 - (T/T_\mathrm{c})^{3/2}\right]} . \tag{10.88}$$

That is, the chemical potential is of the order of $k_\mathrm{B}T/N$. Therefore, for any positive-energy single-particle state, it is justifiable to replace μ by 0.

Let us estimate T_c for the case of helium. Helium gas liquefies at $T = 4.2\,\mathrm{K}$ under atmospheric pressure. For simplicity, let us assume that the gas satisfies the equation of state of an ideal gas down to this temperature. In this case we obtain $N/V = P/k_\mathrm{B}T = 1.74 \times 10^{27}\,\mathrm{m}^{-3}$ at $P = 1.01 \times 10^5\,\mathrm{Pa}$ and $T = 4.2\,\mathrm{K}$. Using this value, we obtain $T_\mathrm{c} = 0.58\,\mathrm{K}$. Thus we cannot expect Bose–Einstein condensation of helium gas at ambient pressure. Next, let us consider liquid helium, which has a molar volume $V/N = 27.6\,\mathrm{cm}^3\,\mathrm{mol}^{-1}$. This gives $T_\mathrm{c} = 3.13\,\mathrm{K}$. Therefore, if the interaction between helium atoms in the liquid phase is negligible, we can expect to have Bose–Einstein condensation at this temperature. In fact, the interaction is not negligible, but liquid helium undergoes a phase transition to the superfluid phase at $T_\lambda = 2.19\,\mathrm{K}$. This is the Bose–Einstein condensation transition that occurs in the presence of interaction between atoms.

For a long time it was thought that Bose–Einstein condensation could not be realized in any substance other than helium. However, in 1995, it became possible to cool vapors of alkali metals, such as Li, Na, and Rb, to below $1\,\mu\mathrm{K}$. The densities of these gases were in the range 10^{18} to $10^{20}\,\mathrm{m}^{-3}$, and at the expected T_c of about 30 nK to 300 nK, evidence of condensation was obtained [13, 14]. Since then, these Bose–Einstein condensates have been intensively investigated. A Nobel Prize was awarded in 2001 to Cornell, Ketterle, and Wieman for the realization of these condensates.

Internal Energy, Pressure, and Heat Capacity

The internal energy can be written as

$$U(T, V, \mu) = \sum_i n_i E_i = \int_0^\infty \mathrm{d}E\, D(E) \frac{E}{\mathrm{e}^{\beta(E-\mu)} - 1}$$
$$= \frac{V}{4\pi^2\hbar^3} (2m)^{3/2} \int_0^\infty \mathrm{d}E\, \frac{E^{3/2}}{\mathrm{e}^{\beta(E-\mu)} - 1} . \tag{10.89}$$

The pressure can be calculated from the partition function J, and related to the internal energy:

$$
\begin{aligned}
P(T,V,\mu)V &= -J = -k_{\mathrm{B}}T\sum_i \ln\left[1-\mathrm{e}^{-\beta(E_i-\mu)}\right] \\
&= -k_{\mathrm{B}}T\int_0^\infty \mathrm{d}E\, D(E)\ln\left[1-\mathrm{e}^{-\beta(E-\mu)}\right] \\
&= \frac{2}{3}U(T,V,\mu)\,.
\end{aligned}
\tag{10.90}
$$

This relation is easily verified by partial integration, just as in the case of the Fermi gas. In fact, the relation $PV = (2/3)U$ is a property common to all noninteracting gases.

To proceed, we need to consider the cases of $T < T_{\mathrm{c}}$ and $T > T_{\mathrm{c}}$ separately. In a Bose–Einstein condensed phase at $T < T_{\mathrm{c}}$, the chemical potential is fixed at $\mu = 0$, and so in this region calculation is easy. Since the Bose–Einstein condensate does not contribute to the energy, U can be calculated as follows:

$$
\begin{aligned}
U(T,V,\mu=0) &= \frac{V}{4\pi^2\hbar^3}(2m)^{3/2}\int_0^\infty \mathrm{d}E\,\frac{E^{3/2}}{\mathrm{e}^{\beta E}-1} \\
&= \frac{V}{4\pi^2\hbar^3}(2m)^{3/2}(k_{\mathrm{B}}T)^{5/2}\int_0^\infty \mathrm{d}x\,\frac{x^{3/2}}{\mathrm{e}^{x}-1} \\
&= \frac{V}{4\pi^2\hbar^3}(2m)^{3/2}(k_{\mathrm{B}}T)^{5/2}\,\Gamma\left(\frac{5}{2}\right)\zeta\left(\frac{5}{2}\right)\,.
\end{aligned}
\tag{10.91}
$$

This can be rewritten using the relation between the particle number N and T_{c} given by (10.84),

$$
N = N(T_{\mathrm{c}},V,\mu=0) = \frac{V}{4\pi^2\hbar^3}(2m)^{3/2}(k_{\mathrm{B}}T)^{3/2}\,\Gamma\left(\frac{3}{2}\right)\zeta\left(\frac{3}{2}\right)\,,
$$

as

$$
\begin{aligned}
U(T,V,N) &= \frac{3}{2}\frac{\zeta(5/2)}{\zeta(3/2)}Nk_{\mathrm{B}}T\left(\frac{T}{T_{\mathrm{c}}}\right)^{3/2} \\
&= 0.7703Nk_{\mathrm{B}}T\left(\frac{T}{T_{\mathrm{c}}}\right)^{3/2}\,.
\end{aligned}
\tag{10.92}
$$

The pressure is then given by

$$
\begin{aligned}
P(T,V,N)V &= \frac{2}{3}U(T,V,N) = \frac{\zeta(5/2)}{\zeta(3/2)}Nk_{\mathrm{B}}T\left(\frac{T}{T_{\mathrm{c}}}\right)^{3/2} \\
&= 0.5135Nk_{\mathrm{B}}T\left(\frac{T}{T_{\mathrm{c}}}\right)^{3/2}\,.
\end{aligned}
\tag{10.93}
$$

Thus, at $T = T_c$, the pressure is reduced to about one-half of the classical value given by $PV = Nk_BT$. From this equation, it might be thought that Boyle's law, $PV =$ constant at fixed temperature, is obeyed. However, this is not correct. Since $T_c \propto (N/V)^{2/3}$, the pressure at $T \leq T_c$ is independent of the volume and depends only on T:

$$P(T, V, N) = k_BT \left(\frac{mk_BT}{2\pi\hbar^2}\right)^{3/2} \zeta\left(\frac{5}{2}\right) . \tag{10.94}$$

From the internal energy, the heat capacity at constant volume can be calculated:

$$\begin{aligned} C_V &= \left(\frac{\partial U(T, V, N)}{\partial T}\right)_{V,N} \\ &= \frac{15\zeta(5/2)}{4\zeta(3/2)} Nk_B \left(\frac{T}{T_c}\right)^{3/2} = 1.926 Nk_B \left(\frac{T}{T_c}\right)^{3/2} . \end{aligned} \tag{10.95}$$

That is, it decreases in proportion to $T^{3/2}$.

Thermodynamic Variables at $T > T_c$

At $T > T_c$, calculation becomes difficult. In order to proceed, we introduce a function $F_\sigma(\alpha)$, where

$$F_\sigma(x) \equiv \frac{1}{\Gamma(\upsilon)} \int_0^\infty \mathrm{d}y \, \frac{y^{\sigma-1}}{\mathrm{e}^{x+y} - 1} = \sum_{n=1}^\infty n^{-\sigma} \mathrm{e}^{-nx} . \tag{10.96}$$

For $\sigma > 1$, the value at $x = 0$ is given by the zeta function: $F_\sigma(0) = \zeta(\sigma)$ $(\sigma > 1)$. The derivative with respect to x is related to $F_{\sigma-1}$, i.e.

$$\frac{\mathrm{d}}{\mathrm{d}x} F_\sigma(x) = -F_{\sigma-1}(x) . \tag{10.97}$$

In terms of this function, the thermodynamic functions can be written as

$$N(T, V, \mu) = \frac{V}{4\pi^2\hbar^3} (2mk_BT)^{3/2} F_{3/2}(-\beta\mu) , \tag{10.98}$$

$$\begin{aligned} U(T, V, \mu) &= \frac{V}{4\pi^2\hbar^3} (2mk_BT)^{3/2} k_BT F_{5/2}(-\beta\mu) \\ &= \frac{3}{2} Nk_BT \frac{F_{5/2}(-\beta\mu)}{F_{3/2}(-\beta\mu)} , \end{aligned} \tag{10.99}$$

and

$$P(T,V,\mu)V = Nk_{\mathrm{B}}T\frac{F_{5/2}(-\beta\mu)}{F_{3/2}(-\beta\mu)}\,. \tag{10.100}$$

Using the expression for T_{c}, we can rewrite (10.98) as

$$N(T,V,\mu) = N\left(\frac{T}{T_{\mathrm{c}}}\right)^{3/2}\frac{F_{3/2}(-\beta\mu)}{\zeta\,(3/2)}\,. \tag{10.101}$$

This equation can be used to determine the chemical potential as a function of T/T_{c}. Namely, μ is determined by solving the equation

$$F_{3/2}(-\beta\mu) = \zeta\left(\frac{T_{\mathrm{c}}}{T}\right)^{3/2}\,. \tag{10.102}$$

An analytical solution of (10.102) cannot be obtained. Thus, here we shall show only that the heat capacity is continuous at T_{c}. We first calculate the derivative of μ using (10.102). As T_{c} is given by the ratio N/V, the derivative at fixed V and N gives

$$F'_{3/2}(-\beta\mu)\left(\frac{\partial(-\beta\mu)}{\partial T}\right)_{V,N} = -\frac{3}{2T}F_{3/2}(-\beta\mu)\,. \tag{10.103}$$

This can be rewritten as follows:

$$F_{1/2}(-\beta\mu)\left(\frac{\partial(-\beta\mu)}{\partial T}\right)_{V,N} = \frac{3}{2T}F_{3/2}(-\beta\mu)\,. \tag{10.104}$$

Now we can calculate the heat capacity:

$$\begin{aligned} C_V &= \left(\frac{\partial U}{\partial T}\right)_{V,N} \\ &= \frac{3}{2}Nk_{\mathrm{B}}\left(\frac{F_{5/2}}{F_{3/2}} + T\frac{F'_{5/2}F_{3/2} - F_{5/2}F'_{3/2}}{F_{3/2}^2}\left(\frac{\partial(-\beta\mu)}{\partial T}\right)_{V,N}\right) \\ &= \frac{3}{2}Nk_{\mathrm{B}}\left(\frac{F_{5/2}}{F_{3/2}} - \frac{3}{2}\frac{F_{3/2}^2 - F_{5/2}F_{1/2}}{F_{3/2}F_{1/2}}\right) \\ &= \frac{3}{2}Nk_{\mathrm{B}}\left(\frac{5}{2}\frac{F_{5/2}}{F_{3/2}} - \frac{3}{2}\frac{F_{3/2}}{F_{1/2}}\right)\,. \end{aligned} \tag{10.105}$$

In this equation, the argument of F_σ, $-\beta\mu$ has been omitted for simplicity. When the temperature approaches T_{c}, the chemical potential vanishes. In this limit, $F_{1/2}$ diverges, and $F_\sigma \to \zeta(\sigma)$ for $\sigma = 3/2$ and $5/2$.[8] Thus, the second term of the heat capacity vanishes at $T = T_\mathrm{c}$, and the first term coincides with the value for the Bose–Einstein condensed phase, (10.95).

In Figs. 10.8–10.10, we show the temperature dependence of various variables: the chemical potential (solid line) and the fraction in the Bose–Einstein condensate (dashed line) are shown in Fig. 10.8, the pressure is shown in Fig. 10.9, and the heat capacity is shown in Fig. 10.10.

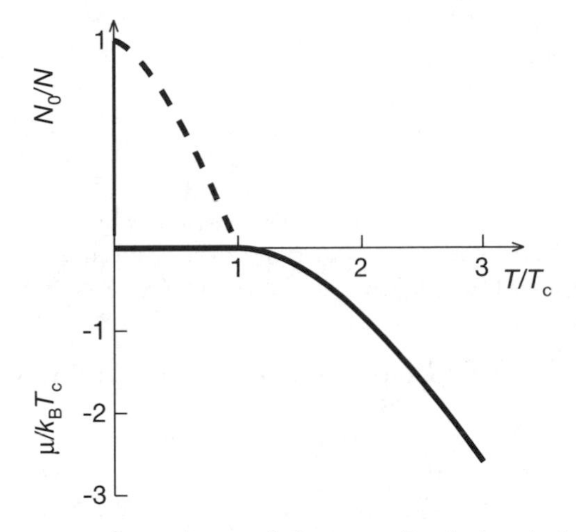

Fig. 10.8. Temperature dependence of the normalized chemical potential $\mu/k_\mathrm{B}T_\mathrm{c}$ and the fraction in the Bose–Einstein condensate, shown by *solid* and *dashed lines*, respectively. The *horizontal axis* shows the reduced temperature T/T_c

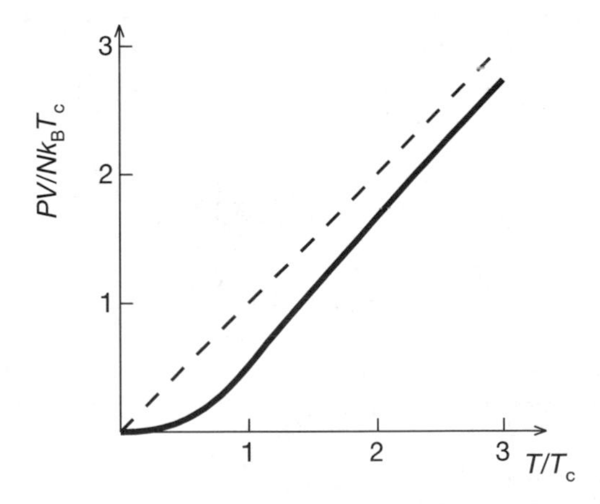

Fig. 10.9. Temperature dependence of $PV/Nk_\mathrm{B}T_\mathrm{c}$, plotted as a function of the reduced temperature T/T_c. The *thin dashed line* shows the classical value, given by $PV/Nk_\mathrm{B}T_\mathrm{c} = T/T_\mathrm{c}$

[8] $F_\sigma(0) = \zeta(\sigma)$ is true only for $\sigma > 1$. When $\sigma = 1/2$, $\zeta(1/2) = -1.4603545$ is finite, but $F_{1/2}(0) = \infty$.

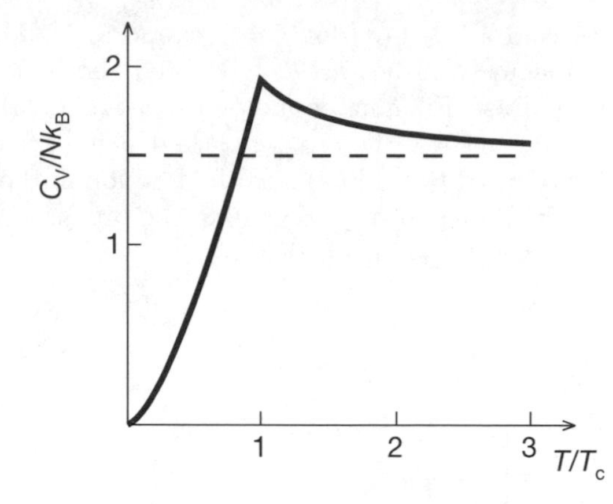

Fig. 10.10. Temperature dependence of the constant-volume heat capacity normalized by Nk_B, C_V/Nk_B, plotted as a function of the reduced temperature T/T_c. The *thin dashed line* shows the classical value, given by $C_V/Nk_B = 3/2$

10.6 Properties of Gases at High Temperature

From Figs. 10.5–10.10, we can see that both the Fermi gas and the Bose gas approach the classical ideal gas at $T \gg T_F$ or $T \gg T_c$. Here we discuss this crossover from a quantum gas to a classical gas. We have seen that a typical Fermi gas, namely the system of conduction electrons in a metal, has a high Fermi temperature, while the Bose–Einstein condensation temperature of a typical Bose gas, namely liquid helium or an alkali metal vapor, is low. However, this difference arises from the difference in the mass of the particles and in the number density. In fact, if the mass and the density were the same, these temperatures would be of the same order. These temperatures are given by

$$k_B T_F = 9.57 \frac{\hbar^2}{2m} \left(\frac{N}{V} \right)^{2/3} \tag{10.106}$$

and

$$k_B T_c = 6.63 \frac{\hbar^2}{2m} \left(\frac{N}{V} \right)^{2/3} . \tag{10.107}$$

These are of the order of the kinetic energy of a particle with a wave number $(N/V)^{1/3}$, or a particle with a wavelength of the order of the mean interparticle distance.

Now let us see what happens when the temperature is higher than these temperatures. As shown in the figures, the chemical potential decreases as the temperature is increased. This behavior is easily understood. If the chemical potential is fixed and the temperature is raised, the occupation probability of each single-particle state increases and the total number of particles N increases. Thus, to keep N fixed, μ must decrease. Since a higher temperature causes higher-energy single-particle states of the system to be populated, the decrease of the chemical potential must occur in such a way that the occupation probability of each single-particle state is decreased. That is, for a fixed energy, the exponential in the denominator of the distribution function $\mathrm{e}^{\beta(E-\mu)}$ must increase as the temperature is increased. Therefore, at sufficiently high temperatures, the condition $\mathrm{e}^{\beta(E-\mu)} \gg 1$ is satisfied, and the distribution functions of both a Fermi and a Bose system can be approximated as follows:

$$\langle n(E)\rangle = \frac{1}{\mathrm{e}^{\beta(E-\mu)} \pm 1} \simeq \mathrm{e}^{-\beta(E-\mu)} . \tag{10.108}$$

In this limit, both Fermi and Bose gases behave like a classical ideal gas. Let us determine the chemical potential in this limit, and calculate $U(T,V,N)$ and $P(T,V,N)$. For simplicity, we consider spinless particles here.[9] The total number of particles at a given chemical potential is

$$\begin{aligned} N &= \int_0^\infty \mathrm{d}E\, D(E)\langle n(E)\rangle \\ &\simeq \frac{V}{4\pi^2}(2m)^{3/2}\int_0^\infty \mathrm{d}E\,\sqrt{E}\,\mathrm{e}^{-\beta(E-\mu)} \\ &= V\left(\frac{mk_\mathrm{B}T}{2\pi\hbar^2}\right)^{3/2}\mathrm{e}^{\beta\mu} . \end{aligned} \tag{10.109}$$

Thus,

$$\mathrm{e}^{-\beta\mu} = \frac{V}{N}\left(\frac{mk_\mathrm{B}T}{2\pi\hbar^2}\right)^{3/2} = \frac{1}{\zeta(3/2)}\left(\frac{T}{T_\mathrm{c}}\right)^{3/2} , \tag{10.110}$$

where the definition of the Bose–Einstein condensation temperature has been used. Therefore, the assumption $\mathrm{e}^{\beta(E-\mu)} \geq \mathrm{e}^{-\beta\mu} \gg 1$ is justified when $T \gg T_\mathrm{c}$. The way in which the chemical potential approaches the classical value as the temperature is raised is shown in Fig. 10.11.

Now that the chemical potential has been determined, we can calculate the internal energy:

$$\begin{aligned} U &= \int_0^\infty \mathrm{d}E\, D(E)E\langle n(E)\rangle \\ &\simeq \frac{V}{4\pi^2}(2m)^{3/2}\int_0^\infty \mathrm{d}E\,(E)^{3/2}\,\mathrm{e}^{-\beta(E-\mu)} \\ &= \frac{V}{4\pi^2}(2m)^{3/2}\,\mathrm{e}^{-\beta\mu}\,\Gamma\left(\frac{5}{2}\right)\frac{1}{\beta^{5/2}} \\ &= \frac{3}{2}Nk_\mathrm{B}T . \end{aligned} \tag{10.111}$$

[9] For a spinless or spin-polarized Fermi gas, the Fermi temperature is given by $T_\mathrm{F} = (\hbar^2/2mk_\mathrm{B})(6\pi^2 N/V)^{2/3} = 15.19(\hbar^2/2m)(N/V)^{2/3}$.

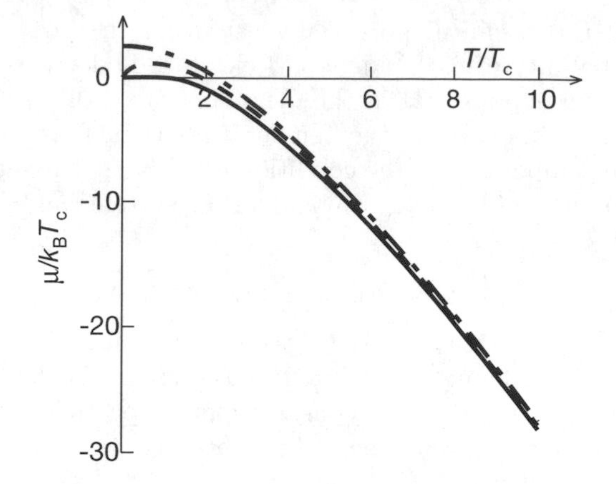

Fig. 10.11. The chemical potentials of Bose and Fermi gases are shown by the *solid line* and *dash-dotted line*, respectively, and that of a classical ideal gas obtained from (10.110) is shown by the *dashed line.* The temperature has been scaled by the Bose–Einstein condensation temperature T_c. The Fermi temperature is given by $T_F = 2.293T_c$

This is the correct result for a classical ideal gas. The pressure can be seen to satisfy the Boyle–Charles equation, $PV = Nk_BT$, from this internal energy, since $PV = (2/3)U$ is satisfied. This result can also be obtained from the partition function J:

$$\begin{aligned} J = -PV &= -k_BT \sum_i \ln\left(1 - e^{-\beta(E_i-\mu)}\right) \\ &\simeq -k_BT \sum_i e^{-\beta(E_i-\mu)} = -k_BT \sum_i \langle n_i \rangle \\ &= -Nk_BT\,. \end{aligned} \tag{10.112}$$

The condition under which the treatment of a Fermi or Bose gas as a classical gas is valid, namely $e^{-\beta\mu} \gg 1$ or $\langle n \rangle \ll 1$ for any single-particle state, is understandable. We discussed at the beginning of this chapter how the reduction of the phase space by a factor of $N!$ becomes invalid when the possibility of occupancy of the same state by more than one particle becomes important. For $\langle n \rangle \ll 1$, such a possibility is negligible, and the classical approximation becomes valid.

It is instructive to derive this condition in other ways. At a temperature T, the particles have an average energy of about k_BT, and so the magnitude of the average momentum will be $p_T = \sqrt{2mk_BT}$. The volume of the single-particle phase space within this momentum p_T is $(4\pi/3)p_T^3V$, and the number of single-particle states in this region of the phase space is $(4\pi/3)p_T^3V/h^3$. The classical approximation should be valid if this number is much larger than N.

The condition for this to apply is

$$\frac{(4\pi/3)p_{\mathrm{T}}^3 V}{h^3} \gg N\,, \tag{10.113}$$

or

$$\frac{4\pi}{3}\frac{V}{N}\left(\frac{mk_{\mathrm{B}}T}{h^2}\right)^{3/2} \gg 1\,. \tag{10.114}$$

This is the same condition as before, except for an unimportant numerical factor. The Planck constant divided by p_{T} is the wavelength of a particle with momentum p_{T}, and is called the thermal de Broglie wavelength and given by

$$\lambda_{\mathrm{T}} = \frac{h}{\sqrt{2mk_{\mathrm{B}}T}}\,. \tag{10.115}$$

The condition above can be written as

$$\left(\frac{4\pi}{3}\frac{V}{N}\right)^{1/3} \gg \lambda_{\mathrm{T}}\,. \tag{10.116}$$

Since the left-hand side gives the mean distance between the particles, this condition can be interpreted as follows. As momenta up to p_{T} are available, the wave functions of the particles can be constructed as wave packets of size λ_{T}. When the interparticle distance is much larger than the size of these wave packets, there is no interference between waves, and no quantum mechanical effects appear. Once the wave packets overlap, however, interference between the waves leads to quantum mechanical effects, and the gas must be treated as a Fermi gas or Bose gas.

I started this book with an introduction to the principles of statistical physics based on consideration of a classical ideal gas. Now we have learned how statistical physics can be applied to various other systems; we have also learned that an ideal gas should actually be described either as a Fermi gas or as a Bose gas, and we have clarified the question of when an approximate treatment of an ideal gas as a classical system is allowed. More topics exist in the field of thermal-equilibrium statistical physics, to which the discussion in this book has been restricted. However, I think that this is an appropriate point at which to end this book on elementary statistical physics. For those who want to learn about more advanced concepts and topics, I recommend several standard books on equilibrium statistical physics [15, 16, 17, 18].

Exercise 17. Estimate the Fermi temperature of copper. The relative atomic mass of copper is 63.546, the density is $8.93\times10^3\,\mathrm{kg\,m^{-3}}$, and the valence is 2. For the mass of a conduction electron, use the value for an electron in vacuum.

Exercise 18. Derive (10.25).

Part IV

Appendices

A

Formulas Related to the Factorial Function

A.1 Binomial Coefficients and Binomial Theorem

If we have a set of N identical particles, there are

$$ {}_N C_n = \frac{N!}{(N-n)!n!} \tag{A.1} $$

ways of choosing n particles from the set; these ways are called combinations, and ${}_N C_n$ is called a binomial coefficient. The following equation can be proved:

$$ (p+q)^N = \sum_{n=0}^{N} {}_N C_n \, p^n q^{N-n} \,, \tag{A.2} $$

which is known as the binomial theorem.

A.2 Stirling's Formula

The factorial of a large integer N can be approximated as follows:

$$ N! \simeq \sqrt{2\pi N} N^N \, \mathrm{e}^{-N} \,. \tag{A.3} $$

This is called Stirling's formula. For example, for $N = 50$, the right-hand and left-hand sides of (A.3) are 3.041×10^{64} and 3.036×10^{64}, respectively. Taking the logarithm of both sides, we obtain

$$ \ln N! \simeq \left(N + \frac{1}{2} \right) \ln N - N + 0.919 \simeq N \ln N - N \,. \tag{A.4} $$

A.3 $n!!$

The notation $n!!$ does not mean the factorial of $n!$. When n is an odd integer,

$$n!! = n \times (n-2) \times (n-4) \times \cdots \times 3 \times 1\,, \tag{A.5}$$

and when n is even,

$$n!! = n \times (n-2) \times (n-4) \times \cdots \times 4 \times 2 = 2^{n/2}(n/2)!\,. \tag{A.6}$$

B

The Gaussian Distribution Function

B.1 The Central Limit Theorem

Consider a situation in which a variable x is realized randomly with some probability. If the probability of observing a value of this variable between x and $x + \mathrm{d}x$ is given by

$$P(x)\,\mathrm{d}x = \frac{1}{\sqrt{2\pi\sigma^2}} \exp\left[-\frac{(x-\mu)^2}{2\sigma^2}\right] \mathrm{d}x\,, \tag{B.1}$$

it is said that the probability distribution function of x is a Gaussian distribution function, or normal distribution function. If x is observed many times, the average is μ, and σ^2 gives the variance; using a Gaussian integral of the kind described below, we can see that

$$\langle x \rangle \equiv \int_{-\infty}^{\infty} \mathrm{d}x\, x P(x) = \mu \tag{B.2}$$

and

$$\langle (x-\mu)^2 \rangle \equiv \int_{-\infty}^{\infty} \mathrm{d}x\, (x-\mu)^2 P(x) = \sigma^2\,. \tag{B.3}$$

Gaussian distributions often appear in statistical physics. The reason arises from the central limit theorem. This theorem says that if there are N random variables x_n $(n = 1, 2, 3, \cdots, N)$, whose distribution functions need not be Gaussian, the probability distribution function of the sum of these variables

$$X \equiv \sum_{n=1}^{N} x_n \tag{B.4}$$

becomes Gaussian in the limit $N \to \infty$. Since many thermodynamic variables, such as the internal energy, pressure, free energy, and magnetization, are given by a summation over contributions from a macroscopic number of particles of the order of the Avogadro number, it is expected that the probability distribution functions of these macroscopic variables will be given by a Gaussian distribution function.

Let us state the theorem more precisely. Let us write the average and the variance of each random variable x_n as $\langle x_n \rangle = e_n$ and $\langle (x_n - e_n)^2 \rangle = v_n$, respectively ($n = 1, 2, \cdots, N$). We write the sums of the averages and of the variances as

$$E_N \equiv \sum_{n=1}^{N} e_n \tag{B.5}$$

and

$$V_N \equiv \sum_{n=1}^{N} v_n \,. \tag{B.6}$$

Then, if $V_N \to \infty$ and $v_n/V_N \to 0$ as $N \to \infty$ for all n, the probability distribution of

$$X \equiv \sum_{n=1}^{N} x_n \tag{B.7}$$

is given by a Gaussian distribution function with an average $\mu = E_N$ and a variance $\sigma^2 = V_N$ as $N \to \infty$:

$$P(X) = \frac{1}{\sqrt{2\pi V_N}} \exp\left[-\frac{(X - E_N)^2}{2V_N} \right] . \tag{B.8}$$

B.1.1 Example

As a simple example, let us consider a situation in which each random variable x_n takes values of ± 1 with equal probability. This is the situation that exists in our toy model of rubber under zero applied force (Chap. 6), in the Ising model in the case of the paramagnetic phase (Chap. 7), and in a one-dimensional random walk, where a person walks back and forth randomly (not treated in this book). In this case the average e_n is equal to 0, and the variance v_n ie equal to 1. Therefore, the probability distribution function of $X \equiv \sum x_n$, $P(X)$, tends to

$$P(X) = \frac{1}{\sqrt{2\pi N}} \exp\left[-\frac{X^2}{2N} \right] \tag{B.9}$$

as $N \to \infty$.

B.2 Gaussian Integrals

Integrals of the following form are called Gaussian integrals, and appear in various situations in physics:

$$\int_{-\infty}^{\infty} \mathrm{d}x \, x^{2n} \mathrm{e}^{-\sigma x^2} = \frac{(2n-1)!!}{2^n} \sqrt{\frac{\pi}{\sigma^{2n+1}}} \,. \tag{B.10}$$

Here n is an integer. In particular, the Gaussian integrals for $n = 0$ and $n = 1$ are

$$\int_{-\infty}^{\infty} \mathrm{d}x \, \mathrm{e}^{-\sigma x^2} = \sqrt{\frac{\pi}{\sigma}} \tag{B.11}$$

and

$$\int_{-\infty}^{\infty} \mathrm{d}x \, x^2 \mathrm{e}^{-\sigma x^2} = \frac{\sqrt{\pi}}{2\sigma^{3/2}} \, . \tag{B.12}$$

Equation (B.11) can be proved in the following way. First we consider the following double integral:

$$\int_{\infty}^{\infty} \mathrm{d}x \int_{-\infty}^{\infty} \mathrm{d}y \, \mathrm{e}^{-\sigma(x^2+y^2)} \, . \tag{B.13}$$

The integrals with respect to x and y are independent, and so we can write this equation as

$$\int_{-\infty}^{\infty} \mathrm{d}x \, \mathrm{e}^{-\sigma x^2} \int_{-\infty}^{\infty} \mathrm{d}y \, \mathrm{e}^{-\sigma y^2} \, . \tag{B.14}$$

This means that this double integral is equal to the square of the left-hand side of (B.11). On the other hand, we can consider the double integral as a surface integral in the xy plane. Using polar coordinates (r, θ) and noting that $x^2 + y^2 = r^2$ and $\mathrm{d}x \, \mathrm{d}y = r \, \mathrm{d}r \, \mathrm{d}\theta$, we can rewrite the double integral in the following form:

$$\int_{0}^{\infty} \mathrm{d}r \int_{0}^{2\pi} \mathrm{d}\theta \, r \mathrm{e}^{-\sigma r^2} \, . \tag{B.15}$$

This integral is easily evaluated, and the result is π/σ. Thus (B.11) has been proved. Equation (B.10) can be derived from (B.11) by differentiating both sides n times with respect to σ.

B.3 The Fourier Transform of a Gaussian Distribution Function

The Fourier transform of a Gaussian distribution function also has the form of a Gaussian distribution function. To show this, we consider the following integral:

$$\begin{aligned} \int_{-\infty}^{\infty} \mathrm{d}x \, \mathrm{e}^{kx} \mathrm{e}^{-\sigma x^2} &= \int_{-\infty}^{\infty} \mathrm{d}x \, \exp\left[-\sigma \left(x - \frac{k}{2\sigma}\right)^2 + \frac{k^2}{4\sigma}\right] \\ &= \int_{-\infty}^{\infty} \mathrm{d}x \, \exp\left[-\sigma x^2 + \frac{k^2}{4\sigma}\right] = \sqrt{\frac{\pi}{\sigma}} \mathrm{e}^{k^2/4\sigma} \, . \end{aligned} \tag{B.16}$$

In the above equation, we have used the fact that we can replace $(x - k/2\sigma)$ by x, as the limits of the integral extend to infinity. This integral can also be evaluated by expanding e^{kx} in a Taylor series as follows:

$$\begin{aligned}\int_{-\infty}^{\infty} \mathrm{d}x\, \mathrm{e}^{kx} \mathrm{e}^{-\sigma x^2} &= \sum_{n=0}^{\infty} \frac{k^n}{n!} \int_{-\infty}^{\infty} \mathrm{d}x\, x^n \mathrm{e}^{-\sigma x^2} = \sum_{n=0}^{\infty} \frac{k^{2n}}{(2n)!} \int_{-\infty}^{\infty} \mathrm{d}x\, x^{2n} \mathrm{e}^{-\sigma x^2} \\ &= \sqrt{\frac{\pi}{\sigma}} \sum_{n=0}^{\infty} \frac{(2n-1)!!}{(2n)!} \frac{k^{2n}}{2^n \sigma^2} = \sqrt{\frac{\pi}{\sigma}} \sum_{n=0}^{\infty} \frac{1}{(n)!} \left(\frac{k^2}{4\sigma} \right)^2 \\ &= \sqrt{\frac{\pi}{\sigma}} \mathrm{e}^{k^2/4\sigma} \,. \end{aligned} \tag{B.17}$$

The Fourier transform of a Gaussian distribution function is obtained by replacing k in these equations by $\mathrm{i}k$:

$$\frac{1}{\sqrt{2\pi}} \int_{-\infty}^{\infty} \mathrm{d}x\, \mathrm{e}^{\mathrm{i}kx} \mathrm{e}^{-\sigma x^2} = \sqrt{\frac{1}{2\sigma}} \mathrm{e}^{-k^2/4\sigma} \,. \tag{B.18}$$

Equation (B.17) tells us that it is permissible to replace $(x - k/2\sigma)$ by x in (B.16), even if k is a complex number. Those who are familiar with complex integrals will be able to prove the validity of this replacement directly using Riemann's theorem. The Fourier transform of a Gaussian distribution function does not appear in this book. However, it appears in various situations in physics, including situations in quantum mechanics and advanced statistical physics.

C

Lagrange's Method of Undetermined Multipliers

This is a method for finding an extremum of $f(x, y)$ under the condition that $g(x, y) = c$, where c is a given constant. Suppose that the equation $g(x, y) = c$ has been solved to obtain $y = h(x, c)$. The problem then reduces to finding an extremum of $f(x, h(x, c))$, and so the condition is

$$\frac{\mathrm{d}}{\mathrm{d}x} f(x, h(x, c)) = \frac{\partial f(x, h)}{\partial x} + \frac{\partial f(x, y)}{\partial y} \frac{\mathrm{d}h(x, c)}{\mathrm{d}x} = 0 \,. \tag{C.1}$$

On the other hand, from the derivative of $c = g(x, y) = g(x, h(x, c))$ with respect to x, we obtain

$$0 = \frac{\partial g}{\partial x} + \frac{\partial g}{\partial y} \frac{\mathrm{d}h}{\mathrm{d}x} \,. \tag{C.2}$$

Thus,

$$\frac{\mathrm{d}h}{\mathrm{d}x} = -\frac{\partial g / \partial x}{\partial g / \partial y} \,. \tag{C.3}$$

We can use this equation to eliminate $\mathrm{d}h/\mathrm{d}x$ in (C.1), and obtain

$$\frac{\partial f(x, h)}{\partial x} - \frac{\partial g / \partial x}{\partial g / \partial y} \frac{\partial f}{\partial y} = 0 \tag{C.4}$$

as the condition for an extremum.

However, the equation $g(x, y) = c$ is usually difficult to solve. In such a case we can try to find an extremum of $f(x, y) - \lambda g(x, y)$ with respect to x and y, where we have introduced an undetermined multiplier λ. We obtain two conditions:

$$\frac{\partial f}{\partial x} - \lambda \frac{\partial g}{\partial x} = 0, \quad \frac{\partial f}{\partial y} - \lambda \frac{\partial g}{\partial y} = 0 \,. \tag{C.5}$$

By eliminating λ from these equations, we obtain the following equation:

$$\frac{\partial f}{\partial x} - \frac{\partial g / \partial x}{\partial g / \partial y} \frac{\partial f}{\partial y} = 0 \,. \tag{C.6}$$

This is the same equation as (C.4). This agreement gives us a strategy for finding an extremum of f under the condition $g = c$. We solve the coupled equations (C.5), and find $(x(\lambda), y(\lambda))$ at which f has an extremum, where λ is a parameter. Then we choose the value of λ such that $g(x(\lambda), y(\lambda)) = c$ is satisfied. This is Lagrange's method of undetermined multipliers.

C.1 Example

To illustrate the method, we consider an easy example here. Let us find an extremum of the function $f(x, y) = ax + by$ under the condition that $g(x, y) = x^2 + y^2 = 1$. We calculate the following derivatives:

$$0 = \frac{\partial f}{\partial x} - \lambda \frac{\partial g}{\partial x} = a - 2\lambda x\,,$$

$$0 = \frac{\partial f}{\partial y} - \lambda \frac{\partial g}{\partial y} = b - 2\lambda y\,. \qquad \text{(C.7)}$$

From these equations, we obtain the values of x and y at which the function f has an extremum: $x = a/(2\lambda)$ and $y = b/(2\lambda)$. The undetermined multiplier λ is determined from the condition that $g = 1$. Namely, from

$$g\left(\frac{a}{2\lambda}, \frac{b}{2\lambda}\right) = \frac{a^2 + b^2}{4\lambda^2} = 1\,, \qquad \text{(C.8)}$$

we obtain $\lambda = \pm\sqrt{a^2 + b^2}/2$. Thus, we find that under the condition $x^2 + y^2 = 1$, the function $f(x, y) = ax + by$ has extremal values of $\pm\sqrt{a^2 + b^2}$ at $x = \pm a/\sqrt{a^2 + b^2}$ and $y = \pm b/\sqrt{a^2 + b^2}$.

A slightly harder example may be obtained by changing the function f to $f = xy + ax + by$. The solution is left as an exercise for the reader.

C.2 Generalization

Up to this point, we have considered the case of two variables x and y. The method can easily be generalized to the case of many variables. To find the extrema of a function of n variables $f(x_1, x_2, \cdots, x_n)$ under the condition that $g(x_1, x_2, \cdots, x_n) = c$, we first solve the following coupled equations for x_i:

$$\frac{\partial f}{\partial x_i} - \lambda \frac{\partial g}{\partial x_i} = 0 \quad (i = 1, 2, \cdots, n)\,. \qquad \text{(C.9)}$$

The solution x_i is put into the function g to determine the values of λ that satisfy $g = c$. The extrema of f can then be calculated.

D

Volume of a Hypersphere

We need to know either the surface area or the volume of a hypersphere in $6N$-dimensional phase space to determine the entropy of an ideal gas. This hypersphere is defined by the condition of constant energy. Here we derive its surface area and volume.

A hypersphere in an n-dimensional space is defined by

$$\sum_{i=1}^{n} x_i^2 = r^2 , \tag{D.1}$$

where x_i is the ith coordinate in the n-dimensional space, and r is the radius of the hypersphere. The volume of the hypersphere is proportional to r^n, and so we write it as $V_n(r) = c_n r^n$. A hypersphere is a circle in two-dimensional space and an ordinary sphere in three-dimensional space. Therefore,

$$c_2 = \pi \tag{D.2}$$

and

$$c_3 = \frac{4}{3}\pi . \tag{D.3}$$

The surface area of an n-dimensional hypersphere is given by

$$S_n(r) = n c_n r^{n-1} , \tag{D.4}$$

since the following equation is satisfied:

$$V_n(r) = \int_0^r S_n(r)\,\mathrm{d}r . \tag{D.5}$$

Now let us consider the following integral:

$$\begin{aligned} I &= \int_0^\infty \mathrm{e}^{-ar^2} S_n(r)\,\mathrm{d}r = nc_n \int_0^\infty \mathrm{e}^{-ar^2} r^{n-1}\,\mathrm{d}r \\ &= \frac{1}{2} nc_n \int_0^\infty \mathrm{e}^{-az} z^{n/2-1}\,\mathrm{d}z \\ &= nc_n \frac{1}{2a^{n/2}} \Gamma\left(\frac{n}{2}\right) . \end{aligned} \tag{D.6}$$

In the second line, the variable of integration has been replaced by $z = r^2$. The gamma function Γ in the last line is defined as follows:

$$\int_0^\infty \mathrm{e}^{-at} t^b\,\mathrm{d}t = \frac{\Gamma(b+1)}{a^{b+1}} . \tag{D.7}$$

It can be shown that $\Gamma(x+1) = x\Gamma(x)$ and that $\Gamma(n) = (n-1)!$, by partial integration of the defining equation (D.7). We can also show that $\Gamma(1) = 1$ and $\Gamma(1/2) = \sqrt{\pi}$. On the other hand, the integral I can be interpreted as a volume integral of e^{-ar^2} in the n-dimensional space, and can be rewritten using the relation $r^2 = x_1^2 + x_2^2 + x_3^2 + \cdots + x_n^2$:

$$\begin{aligned} I &= \int_{-\infty}^\infty \mathrm{d}x_1 \int_{-\infty}^\infty \mathrm{d}x_2 \cdots \int_{-\infty}^\infty \mathrm{d}x_n \exp\left\{-a\left(x_1^2 + x_2^2 \cdots + x_n^2\right)\right\} \\ &= \left\{\int_{-\infty}^\infty \exp\left(-ax^2\right)\mathrm{d}x\right\}^n = \left(\sqrt{\frac{\pi}{a}}\right)^n . \end{aligned} \tag{D.8}$$

From (D.6) and (D.8), we obtain

$$c_n = \frac{2\pi^{n/2}}{n\Gamma(n/2)} = \frac{\pi^{n/2}}{\Gamma(n/2+1)} . \tag{D.9}$$

E

Hyperbolic Functions

The functions $\sinh x$, $\cosh x$, $\tanh x$, and their inverses are known as hyperbolic functions. These functions have an intimate relationship to the trigonometric functions. The definitions of the hyperbolic functions are

$$\sinh x = \frac{\mathrm{e}^x - \mathrm{e}^{-x}}{2}, \tag{E.1}$$

$$\cosh x = \frac{\mathrm{e}^x + \mathrm{e}^{-x}}{2}, \tag{E.2}$$

$$\tanh x = \frac{\sinh x}{\cosh x}, \tag{E.3}$$

$$\coth x = \frac{1}{\tanh x}, \tag{E.4}$$

$$\mathrm{sech}\, x = \frac{1}{\cosh x}, \tag{E.5}$$

$$\mathrm{cosech}\, x = \frac{1}{\sinh x}. \tag{E.6}$$

Therefore, when these functions are used with imaginary arguments, they become trigonometric functions:

$$\cosh(\mathrm{i}x) = \cos x, \quad \sinh(\mathrm{i}x) = \mathrm{i}\sin x \tag{E.7}$$

Figure E.1 shows the behavior of these functions.

The hyperbolic functions satisfy the following relations:

$$\begin{aligned} \cosh^2 x - \sinh^2 x &= 1, \\ 1 - \tanh^2 x &= \mathrm{sech}^2 x. \end{aligned} \tag{E.8}$$

Their derivatives are given by

$$\begin{aligned}(\sinh x)' &= \cosh x\,,\\ (\cosh x)' &= \sinh x\,,\\ (\tanh x)' &= \mathrm{sech}^2 x\,,\end{aligned}\tag{E.9}$$

and they have the following series expansions at small x:

$$\begin{aligned}\sinh x &= x + \frac{1}{3!}x^3 + \frac{1}{5!}x^5 + \frac{1}{7!}x^7 + \cdots\,,\\ \cosh x &= 1 + \frac{1}{2!}x^2 + \frac{1}{4!}x^4 + + \frac{1}{6!}x^6 \cdots\,,\\ \tanh x &= x - \frac{1}{3}x^3 + \frac{2}{15}x^5 - \frac{17}{315}x^7 \cdots\,.\end{aligned}\tag{E.10}$$

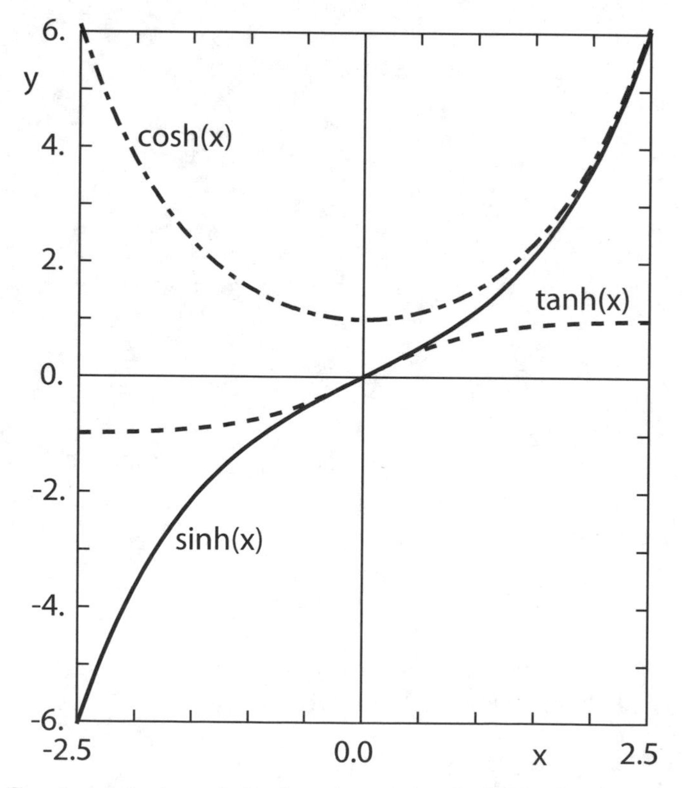

Fig. E.1. Graphs of the hyperbolic functions $\sinh x$ (*solid line*), $\cosh x$ (*dash-dotted line*), and $\tanh x$ (*dashed line*)

F

Boundary Conditions

Waves in a rectangular box can be described by means of a superposition of normal modes. Particles behave as waves, according to quantum mechanics, and so they can also be described by normal modes. A normal mode is characterized by its wave vector or momentum. In this book, the molecules of an ideal gas, considered in Chaps. 4 and 10, and lattice vibrations and electromagnetic waves, considered in Chap. 5, are treated as waves in a box. The wave vector $\boldsymbol{k}$ and momentum $\boldsymbol{p}$ are related by de Broglie's relation $\boldsymbol{p} = \hbar \boldsymbol{k}$, and the allowed values of the wave vector are determined by the boundary condition. Here we summarize the relation between the boundary condition and the allowed values of the wave vector.

F.1 Fixed Boundary Condition

One typical boundary condition is the fixed boundary condition. In this condition, the displacement of the wave must be zero at a wall. For example, a sound wave in air is a longitudinal wave, and the motion of the air perpendicular to the wall must be zero at the boundary. The wave function of an electron or atom also vanishes at a wall.

First, we consider the one-dimensional case. When a wave is confined in a region $0 \leq x \leq L_x$, the wave must have the following form for it to vanish at $x = 0$:

$$\psi(x, t) = A \sin(kx) \cos(\omega t + \alpha) . \tag{F.1}$$

Here A is the amplitude of the wave, and α is the phase. The condition that $\psi(L_x, t) = 0$ at $x = L_x$ must also be fulfilled. As a result, the wave number k must be one of the following k_n, where n is a natural number:

$$k_n = \left(\frac{\pi}{L_x} \right) n . \tag{F.2}$$

Here n must be positive, because a wave with a negative value of k is the same wave as one with a wave number $|k|$. If there is an upper limit on the wave number k so that $k \leq k_{\mathrm{M}}$, the number of allowed values of k is $k_{\mathrm{M}} L_x / \pi$.

This result can be extended to the three-dimensional case. If the wave is confined in a box of dimensions $L_x \times L_y \times L_z$, the allowed wave vectors are given by

$$\boldsymbol{k} = \left(\frac{\pi}{L_x} n_x, \frac{\pi}{L_y} n_y, \frac{\pi}{L_z} n_z \right), \tag{F.3}$$

where n_x, n_y, and n_z are natural numbers.

F.2 Periodic Boundary Condition

In the case of the fixed boundary condition, the existence of the walls is considered explicitly. In this case the situation near the walls is somewhat different from that in the interior, far from the walls. However, there is another boundary condition that can be used, for which we need not consider the existence of the walls explicitly. This is the periodic boundary condition. For a one-dimensional system, we impose the condition that

$$\psi(x + L_x, t) = \psi(x, t). \tag{F.4}$$

A wave satisfying this condition now has the form of a propagating wave instead of the standing wave described by (F.1), namely

$$\psi(x, t) = A \cos(kx - \omega t + \alpha). \tag{F.5}$$

From the boundary condition (F.4), the allowed values of the wave number k_n are given by

$$k_n = \frac{2\pi}{L_x} n, \tag{F.6}$$

where n is an integer. In this case the wave is a propagating wave, and the direction of propagation depends on the sign of n. The interval between adjacent values of k_n in the present case is twice that for (F.2). However, since a negative value of n describes a different wave, the number of waves that satisfy $|k| \leq k_M$ is $k_M L_x / \pi$, as before.

In the three-dimensional case, the allowed wave vectors are given by

$$\boldsymbol{k} = \left(\frac{2\pi}{L_x} n_x, \frac{2\pi}{L_y} n_y, \frac{2\pi}{L_z} n_z \right), \tag{F.7}$$

where n_x, n_y, and n_z are integers.

A periodic boundary condition is difficult to realize experimentally. On the other hand, it is often more convenient for theoretical considerations. When we are dealing with the properties of a bulk sample, the result should be independent of the boundary condition. This is because the contribution from regions near the wall is negligible compared with that from the interior of the sample, which represents most of the volume. In fact, the number of states that satisfy $|k| \leq k_{\mathrm{M}}$ does not depend on the boundary condition. Since the energy of a wave is determined by the frequency of the wave, which depends on the absolute value of the wave vector, the density of states is independent of the boundary condition. Therefore, the properties of a bulk sample calculated by use of statistical mechanics do not depend on the boundary condition.

G

The Riemann Zeta Function

The Riemann zeta function $\zeta(z)$ at $\mathrm{Re}\, z > 1$ is defined by the infinite series:

$$\zeta(z) = \sum_{n=1}^{\infty} \frac{1}{n^z} \quad (\mathrm{Re}\, z > 1)\,. \tag{G.1}$$

The value at $\mathrm{Re}\, z \leq 1$ is given by the analytic continuation of (G.1) to $\mathrm{Re}\, z \leq 1$. This function appears in statistical physics because of the following two integral representations. One of these representations is related to the Bose distribution function:

$$\zeta(z) = \frac{1}{\Gamma(z)} \int_0^{\infty} \frac{t^{z-1}}{\mathrm{e}^t - 1}\, \mathrm{d}t \quad (\mathrm{Re}\, z > 1)\,. \tag{G.2}$$

The other is related to the Fermi distribution function:

$$\zeta(z) = \frac{1}{(1 - 2^{1-z})\,\Gamma(z)} \int_0^{\infty} \frac{t^{z-1}}{\mathrm{e}^t + 1}\, \mathrm{d}t \quad (\mathrm{Re}\, z > 0)\,. \tag{G.3}$$

In these equations, $\Gamma(z)$ is the gamma function, defined in (D.7).

It is easy to verify these representations. For example, (G.2) can be transformed to

$$\begin{aligned}
\zeta(z) &= \frac{1}{\Gamma(z)} \int_0^{\infty} \frac{t^{z-1}}{\mathrm{e}^t} \frac{1}{1 - \mathrm{e}^{-t}}\, \mathrm{d}t \\
&= \frac{1}{\Gamma(z)} \int_0^{\infty} t^{z-1} \sum_{n=0}^{\infty} \mathrm{e}^{-(n+1)t}\, \mathrm{d}t \\
&= \sum_{n=0}^{\infty} \frac{1}{(n+1)^z} \\
&= \sum_{n=1}^{\infty} \frac{1}{n^z}\,.
\end{aligned} \tag{G.4}$$

Equation (G.3) can be verified similarly.

For $z = 2$ and 4, the zeta function can be calculated analytically. The results are

$$\zeta(2) = \frac{\pi^2}{6}\,,$$

$$\zeta(4) = \frac{\pi^4}{90}\,. \tag{G.5}$$

The following asymptotic expansion is convenient for calculating $\zeta(z)$ for an arbitrary value of z:

$$\zeta(z) = \sum_{m=1}^{n} \frac{1}{m^z} + \frac{1}{(z-1)n^{z-1}} - \frac{1}{2n^z} + \frac{z}{12n^{z+1}} - \frac{z(z+1)(z+2)}{720n^{z+3}} + \frac{z(z+1)(z+2)(z+3)(z+4)}{30240n^{z+5}} \cdots\,. \tag{G.6}$$

Using this expansion, we can obtain the following values, used in Sect. 10.5.2 in relation to the ideal Bose gas:

$$\zeta\left(\frac{3}{2}\right) = 2.61237535\,, \tag{G.7}$$

$$\zeta\left(\frac{5}{2}\right) = 1.34148726\,. \tag{G.8}$$

References

1. C. Seife: Science **302**, 2038 (2003).
2. P.G. de Gennes: *Scaling Concepts in Polymer Physics* (Cornell University Press, Ithaca, NY, 1979).
3. M. Doi, S.F. Edwards: *The Theory of Polymer Dynamics* (Oxford University Press, 1986).
4. R. Rosenberg: Phys. Today **58**(12), 50 (2005).
5. D.D. Osheroff, R.C. Richardson, D.M. Lee: Phys. Rev. Lett. **28**, 885 (1972).
6. P.S. Epstein: *Textbook of Thermodynamics* (Wiley, New York 1937).
7. L. Onsager: Phys. Rev. **65**, 117 (1944).
8. I.S. Gradshtein, I.M. Ryzhik: *Table of Integrals, Series and Products* (Academic Press, Orlando, 1980).
9. C.N. Yang: Phys. Rev. **85**, 808 (1952).
10. H.E. Stanley: *Introduction to Phase Transitions and Critical Phenomena* (Oxford University Press, 1997).
11. S.-K. Ma: *Modern Theory of Critical Phenomena* (Addison-Wesley, Reading, MA, 1976).
12. L.D. Landau: Sov. Phys. JETP **3** 920 (1957); Sov. Phys. JETP **8**, 70 (1959).
13. M.H. Anderson, J.R. Ensher, M.R. Matthews, C.E. Wieman, E.A. Cornell: Science **269**, 198 (1995).
14. K.B. Davis, M.-O. Mewes, M.R. Andrews, N.J. van Druten, D.S. Durfee, D.M. Kurn, W. Ketterle: Phys. Rev. Lett. **75**, 3969 (1995).
15. L.D. Landau, E.M. Lifshitz: *Statistical Physics*, Course of Theoretical Physics, Vol. 5 (Pergamon, Oxford, 1980).
16. R. Kubo, H. Ichimura, T. Usui, N. Hashitsume: *Statistical Mechanics*, North-Holland Personal Library (North-Holland, Amsterdam, 1988).
17. G.H. Wannier: *Statistical Physics*, Dover Books on Physics and Chemistry (Dover, New York, 1987).
18. F. Reif: *Fundamentals of Statistical and Thermal Physics*, McGraw-Hill Series in Fundamentals of Physics (McGraw-Hill, New York, 1965).

Index

Printing: Krips bv, Meppel
Binding: Stürtz, Würzburg